Adoption of agricultural innovations by smallholder farmers in the context of HIV/AIDS

Adoption of agricultural innovations by smallholder farmers in the context of HIV/AIDS

The case of tissue-cultured banana in Kenya

AWLAE SERIES No. 7

Faith Njeri Nguthi

Wageningen Academic
Publishers

ISBN 978-90-8686-069-2

First published, 2008

Wageningen Academic Publishers
The Netherlands, 2008

Acknowledgements

This book is based on a PhD thesis that was successfully defended on the 25th of June, 2007, at Wageningen University. I attribute the successful completion of the PhD studies to the support of various individuals and institutions whose contributions I would like to acknowledge. I would like to acknowledge the AWLAE program who offered me the opportunity to study for a PhD at Wageningen University through a competitive scholarship offered by the Netherlands government through the Ministry of Foreign Affairs (DGIS). AWLAE stands for *African Women Leaders in Agriculture and Environment*. I greatly appreciate the financial support which enabled me to successfully complete my studies. I thank the government of Kenya and in particular the Director of KARI for granting me study leave, which enabled me to successfully complete my study.

I would like to thank my academic supervisors at Wageningen University and my colleagues in the AWLAE program, for the moral, academic and sisterly support which they provided during my study period. Special thanks are due to the research team during my fieldwork. I offer my gratitude to all the interviewers for their diligence in carrying out the survey. I would like to thank my field research assistants, Francis Wambugu and Mrs. Njoroge and the extension officers in Maragua division especially Kariuki and Maina for their support. Special thanks to Esther Muthoka for her diligence in data entry and subsequent cleaning up of the data. I also appreciate the support provided by the Director, National Horticultural Research Centre, Thika, Dr. S. Waturu and the staff. I am indebted to the people in the study area without whose co-operation it would have been impossible to carry out the study. I greatly appreciate their willingness and enthusiasm to participate in the research and to volunteer information. Special thanks to all the key informants for offering all the valuable information and time. I deeply appreciate Edward Masafu's artistic gifts and thank him for designing the beautiful cover of the book.

Lastly but not least I would like to express my sincere gratitude to my husband Stephen, our children Nellex, Fiona and Karanja. Stephen, I thank you for the sacrifice you made to take care of our children and home when I was away for all those lonely months.

Faith N. Nguthi
Thika, March 2008

Table of contents

List of acronyms and abbreviations

AFC	Agricultural Finance Co-operation
AIDS	Acquired Immunodeficiency Syndrome
ANC	Antenatal Clinic
ARV	Anti-retroviral
AWLAE	African Women Leaders in Agriculture and Environment
CBO	Community-based Organization,
DAEO	Divisional Agricultural Extension officer
DFID	Department for International Development
DGIS	Dutch Department for Development Co-operation
FGD	Focus group discussion
FHIA	Honduran Foundation for Agricultural Research
FIPS	Farm Input Promotion Service
GDP	Gross domestic Product
HIV	Human Immunodeficiency Virus
IDRC	International Development Research Centre
INIBAP	International Network for the Improvement of Banana and Plantain
ISAAA	The International Services for Acquisition of Agri-biotech Applications
ITSC	Institute for Tropical and Sub-tropical and Crops
KARI	Kenya Agricultural Research Institute
KCC	Kenya Creameries Cooperative
KSh	Kenya Shilling
KHDS	Kenya Health Demographic Survey
MOA	Ministry of Agriculture
MOH	Ministry of Health
NACC	National AIDS Control Council
NGO	Non-governmental Organisation
PLWA	People living with HIV/AIDS
PRA	Participatory Rural Appraisal
RF	Rockefeller Foundation
ROSCA	Rotating Savings and Credit Association
SACCO	Savings and Credit Co-operative
SAP	Structural Adjustment Program
SSA	Sub-Sahara Africa
STI	Sexually transmitted infection
UNAIDS	The joint United Nations Program for HIV/AIDS
US$	United States Dollar
VCT	Voluntary Counselling and Testing Centre
WHO	World Health Organisation

Chapter 1

Introduction

1.1 Justification

The use of improved technologies has remained the major strategy used by governments to increase agricultural productivity and promote food and livelihood security. The Kenya government is no exception in view of the fact that 70-80 per cent of the population earns a living from agriculture. The agricultural sector contributes about 30 per cent of the GDP and accounts for 80 per cent of national employment, mainly in the rural areas. In addition, the sector contributes more than 60 per cent of the total export earnings and about 45 per cent of government revenue, while providing for most of the country's food requirements. The sector is estimated to have a further indirect contribution of nearly 27 per cent of GDP through linkages with manufacturing, distribution, and other service-related sectors. In the first decade after independence agricultural production grew by 4.7 per cent annually (Karanja, 2002). However, this impressive growth rate did not continue in the subsequent decades and today agricultural production has shrunk to an annual growth rate of 1.8 per cent (Republic of Kenya, 2003a).

This situation is aggravated by the HIV/AIDS pandemic. According to UNAIDS (2003), in 2003 HIV prevalence rate in Kenya was estimated at 6.7 per cent and 1.2 million Kenyans were living with HIV/AIDS. Although interrelations between the epidemic and overall development have been acknowledged, until recently the linkages to agriculture have received little attention because the epidemic was perceived as being largely urban. However, this scenario is rapidly changing and the greatest burden is now in the rural areas. The disease is now affecting the farming population, especially people in their most productive years (ages 15 to 45). Morbidity and mortality caused by HIV/AIDS result in severe labour shortages for farm work, loss of income and sale of assets. Thus the current HIV/AIDS pandemic challenge is to develop technologies that meet the needs of affected farming households and to maintain high productivity levels.

Recently, KARI[1] in collaboration with development partners Rockefeller Foundation and the International Development Research Centre–Canada (RF and IDRC), have been involved in the development and dissemination of the tissue-cultured banana technology in order to increase income, food security which will ultimately ensure the livelihood security of small-scale farming households.

---

[1] Kenya Agricultural Research Institute is a government institution charged with conducting agricultural research in the country.

This study aims at assessing the suitability of the tissue-cultured banana technology, given the debilitating effects of HIV/AIDS on farming households' assets and resources in Central Kenya. Using the livelihood approach the study investigates how farming households' assets and resources have been affected by HIV/AIDS and how these effects influence the households' ability to utilize the tissue-cultured banana. For this study, both quantitative and qualitative methods of data collection were used. These included focus group discussions, a large-scale formal survey, in-depth semi-structured interviews and secondary data collection.

This project is part of the AWLAE Program. AWLAE stands for African Women Leaders in Agriculture and the Environment. The project, which is funded by Netherlands Directorate-General for International Cooperation (DGIS), aims at developing a gendered framework for research on livelihood and food systems in a situation of HIV/AIDS in Sub-Sahara Africa. The program is inspired by the fact that women produce between 60 and 80 percent of the food in most developing countries and are responsible for half of the world's food production. Despite their key role as food producers and providers their critical contribution to household food security is only now becoming recognized. In sub-Saharan Africa women's role in food production and household food security is likewise significant. The HIV/AIDS pandemic however affects their capability to carry out this role and adds to their burden because of women's responsibility for care.

1.2 Research problem

Banana is one of the crops that have received increased research attention over the last ten years in Kenya. Previously the crop was considered a semi-subsistence women's crop that provided more or less continuous income flow throughout the year, even under a low input regime. In recent years the crop has become an important commercial crop and Qaim (1999) attributes this increase in commercialization to demand due to increasing urbanization as well as diminishing farm incomes from more traditional cash crops, notably coffee. The collapse of the Coffee Agreement in 1989 (Beets, 1990) resulted in an immediate drop in coffee prices and over the years coffee production in Kenya has experienced several ups and downs (Karanja, 2002). Small-scale farmers who depended heavily on proceedings from coffee for their livelihood have had to look for other sources of income by diversification and commercialisation of traditional food crops such as the banana. The banana has the potential for food and livelihood security as it can both be consumed at home as a staple and sold in the market for cash.

The study area, Maragua district, lies in the main and marginal coffee zones where banana has been the main source of livelihood for the small-scale farmers since the collapse of the coffee sector in the mid-eighties. In 2003 it is estimated that 3355 hectares of arable land in the district was under banana, producing a total of 38,040 tons which was valued at Ksh 285.3 m (MOA, 2004). This was second to tea which earned the highest income value. Despite the economic importance of banana in the area production has also been on the decline in the last ten years mainly because of pests and diseases. To reverse this decline the government through KARI introduced the tissue culture technology by which clean planting material of superior varieties were introduced to farmers. This technology was perceived as having the potential to help reverse the declining banana production trends in the area and thus promote food and livelihood security in banana farming households.

1.2.1 The tissue culture technology

The tissue culture technique (micro-propagation) refers to the production of large numbers of plants from small pieces of the stock plant in relatively short periods of time. Depending on the species in question, the original tissue piece may be taken from shoot tip, leaf, lateral bud, stem or root tissue. Micro-propagation offers several distinct advantages compared to conventional propagation techniques. A single ex-plant can be multiplied into several thousand plants in less than one year. Using the technique, it is possible to rapidly introduce selected superior clones of plants in sufficient quantities to have an impact at the farm level. The technique also allows mass multiplication of species that are difficult to regenerate by conventional methods of propagation and where conventional methods of propagation are inadequate to meet the demand of planting material, which is the case with the banana in Kenya. Conventionally one banana plant produces about ten suckers in a year while over five hundred plantlets can be produced using the tissue culture technique. Another purpose for which plant tissue culture is uniquely suited is in the obtaining, of specific pathogen-free plants. Tissue culture plants thus exhibit significantly increased vigour, yield, and early maturity and are disease and pest free.

Studies have shown that in those parts of sub-Saharan Africa where the HIV/AIDS epidemic has reached peaks household labour quality and quantity are reduced, first when HIV-infected persons fall sick, and later when the supply of household labour declines because of patient care and death (Müller, 2004). For example, in one village in Tanzania, in households with an AIDS patient, nearly 30 per cent of household labour was spent on AIDS-related matters (including care of the patient and funeral duties). If two people were devoted to nursing the patient, as was the case in 66 per cent of recorded cases, the total labour loss was 43 per cent on average. As the cost of death and illnesses rise, there is reduction or liquidation of any savings they may have as well as a reduction in their asset base of equipment and animals. Apart from material loss, rural households are also loosing social solidarity through loosening of social bonds and thinning of social fabrics (Rugalema, 1999).

A compounding factor is that infection rates are higher among women, who account for 70% of the agricultural labour force and 80 per cent of food production. In Kenya, women supply 70-75 per cent of agricultural labour and their contribution has been termed as the backbone of the household production (Spring, 2000). Yet it has been reported that women in Kenya have a higher prevalence rate (9%) than men (5%). Furthermore, new technologies have been known to change labour allocations within farming households. Many studies have shown that women's labour burden increases with new technologies and this coupled with HIV/AIDS responsibility of caring for the sick, may have a potential negative effect on the suitability of introduced technologies.

Since technology adoption is strongly linked to the asset base this study adapted the sustainable livelihood approach which is based on the premise that the asset status of people is fundamental to understanding the options open to them, the strategies they adopt to attain livelihoods, the outcomes they aspire to and the vulnerability context under which they operate (Ellis, 2000). Three main dimensions of the livelihood approach were studied: the assets of the rural households, the processes that influence access to those assets and the livelihood outcomes.

All the five capitals used in the livelihood framework (DFID, 2001) are involved when adopting new technologies. For example, human capital in the form of knowledge and skills is required to properly make use of many new technologies. On the other hand assets are only important to livelihoods if they can be accessed and have an ability to support livelihoods. Access to assets is, in turn affected by the vulnerability context or shocks that the household is exposed to. HIV/AIDS is a shock that destroys assets both directly through its impact on human capital as well as indirectly through the sale and disposal of assets to meet medical costs (Barnett & Whiteside, 2002).

Depending on the assets people have, the structures and processes that influence them and the vulnerability context under which they operate, people engage in livelihood activities and strategies that best provide them with livelihood outcomes. Different authors identify various types of strategies. Scoones (1998) categorises strategies into agricultural intensification and extensification; livelihood diversification that includes both paid employment and rural enterprises; and migration for income generation and remittances. Understanding the diverse and dynamic livelihood strategies and activities is important when introducing new activities and strategies in people's livelihoods. In this study strategies and activities were identified as those related to agricultural production, off-farm and non-farm employment, income generating activities (rural trade, services or manufacture), migration and remittances.

Livelihood outcomes are the achievements and benefits that households hope to derive through the implementation of specific activities and strategies. Potential outcomes include conventional indicators such as income, food security and sustainable use of the natural resources. Outcomes may also include reduced vulnerability, strengthened asset base, and improvements in well-being aspects such as health.

From the foregoing discussion it is assumed that income and food security were among the major livelihood priorities of the farming households in the area of study. Banana farming is only one of the livelihood activities that farming households are engaged in and to enable the farming households pursue this activity effectively and efficiently, they adopt the tissue-culture banana farming technology introduced by the agricultural research and the extension system. This technology has the potential to increase banana production as well as income, thus promoting food and livelihood security. However, these farming households are expected to have been hit by the HIV/AIDS epidemic, which may have ravaged and eroded their assets to the extent that their ability to take up or utilize these improved technologies is reduced. Furthermore gender relations within the farming households, in regard to access to factors of production such as labour, land and capital, influence the decisions made about adopting the tissue culture banana technology.

Apart from the impacts of HIV/AIDS on farming households the epidemic may also have influenced the ability of institutions such as the extension services systems that facilitate and enable the households to access and utilize the improved technologies. Based on these suppositions, this research was carried out to assess the constraints that have been brought about by HIV/AIDS for banana farming households in Central Kenya and the suitability of the tissue culture banana under these conditions. The research also assessed how farming households' livelihood strategies had changed and what new coping strategies had emerged. In doing so gender-specific aspects of these constraints were examined.

1.3 Research objectives and questions

1.3.1 Goals and objectives

The goal of this study is to contribute to the livelihood theory by assessing the relevance of agricultural technologies for rural households who are confronted with HIV/AIDS-induced shocks and stress in generating their livelihood. The tissue-cultured banana will be used as a case study. It is anticipated that findings will be used to contribute to policy formulation in the development and dissemination of agricultural technology in smallholder farming systems where households' livelihoods are challenged by HIV/AIDS.

1.3.2 Objectives

a. To assess assets of banana farming households and how these assets influence their livelihood objectives, options, activities and outcomes.
b. To evaluate the impacts of HIV/AIDS on the farming households' assets.
c. To evaluate the suitability of the tissue-cultured banana technology for HIV/AIDS affected households.
d. To assess the performance of the agro-institutional environment in relation to the introduction and utilisation of the technology, and the impacts of the HIV/AIDS on this.

1.3.3 Research questions and sub-questions

1. How do assets available to farming households influence their livelihood options, activities and outcomes?
   a. What assets do farming households have access to?
   b. How do these assets influence the livelihood options and activities of the households?
   c. To what extent is the access to assets and livelihood options and activities gendered?
2. What are the impacts of HIV/AIDS on farming households' assets and consequently livelihood options, activities and outcomes?
   a. To what extent do the impacts of HIV/AIDS affect farming household assets?
   b. How does the impact of HIV/AIDS on assets influence household livelihood options and activities?
   c. To what extent are these effects gendered?
   d. To what extent does HIV/AIDS influence the performance of the agricultural extension system and how does this affect adoption of tissue-cultured?
3. What role does the tissue-cultured banana play in the livelihood activities and outcomes of farming households?
   a. How do households perceive the compatibility of the tissue-cultured banana technology with the existing farming household livelihood assets, options and activities?
   b. How do household assets influence farming household's ability to adopt tissue- cultured banana?
   c. How do the effects of HIV/AIDS on households' assets, including labour influence the household's ability to adopt the tissue-cultured banana?
   d. To what extent does the adoption of tissue-cultured banana influence the livelihood outcomes of households?

1.4 Structure of the book

In Chapter 2 literature review and theoretical approaches to that explain technology adoption is presented. These include the innovation-diffusion model, the economic constraint model, central source of innovation model and the multiple source of innovation model. The sustainable livelihood approach is discussed and presented as the point of departure. The dimensions of the livelihood approach i.e. livelihood assets, activities strategies, outcomes, institutional environment and the vulnerability context are placed in the context of this. Vulnerability is related to the effects of HIV/AIDS on households' livelihood assets. The concept of household as the unit of analysis is discussed. The gendered impacts of HIV/AIDS and their implication for agriculture are also discussed. At the end of the chapter a conceptual framework is presented which depicts how the concepts are interrelated in answering the research questions.

Chapter 3 starts with a brief description of the study area to show its appropriateness as the area of study. The research strategy and study design, the methods of data collection and analysis are discussed in detail. The chapter concludes with a discussion of ethical considerations and of the problems encountered in data collection.

Chapter 4 presents a brief profile of Kenya and the specific study area of Maragua district. Geographic and demographic characteristics, ethnicity, social structure and religion are presented. The role of agriculture in the economy and the government policy changes over the years are discussed. In the third part, banana production and the introduction of the tissue-culture banana technology is presented. A discussion of the marketing structure and gender issues in banana marketing is presented. The HIV/AIDS situation in Kenya and the interrelations between agriculture, gender and HIV/AIDS are explored and peoples' perceptions of the genesis, spread and impacts of HIV/AIDS in the study area are presented.

In Chapter 5 farm households in the sample population are characterized and classified into three comparatively homogenous categories using the factor and cluster analysis procedures. Subsequently a description of the resulting farm typologies, referred to as low, medium and high resource endowment farming households based on their assets endowment, is presented. This discussion is structured around the five capitals of the DFID livelihood pentagon (DFID, 2001). Finally a discussion of the livelihood strategies and activities of the farm household typologies is presented.

The discussion on the effects of HIV/AIDS on assets and livelihood activities in Chapter 6 is disaggregated into HIV/AIDS-affected and non-affected households.

The factors that determine livelihood activities undertaken in HIV/AIDS-affected are investigated. The direct and indirect impacts of HIV/AIDS on the extension services in the study area are as well discussed.

Chapter 7 focuses on the significance of tissue-cultured banana production in the livelihood activities and outcomes of the banana-cultivating households. The perceptions of the farming households on tissue-cultured plantlets as compared to the conventional suckers, are presented. Household characteristics, farm assets, livelihood activities and institutional factors that determine adoption of tissue-cultured banana are explored.

In Chapter 8, the findings of the study are summarised to answer the research questions formulated in above. After a general discussion, the policy implications of the research findings in the development and dissemination of agricultural technology in smallholder farming systems concludes the chapter.

Chapter 2

Literature review and conceptual framework

This chapter starts with a literature review on theoretical approaches that explain technology adoption. These include the innovation-diffusion model, the economic constraint model, the central source of innovation model and the multiple source of innovation model. Subsequently, the sustainable livelihood approach is discussed and presented as the theoretical basis for the study. The dimensions of the livelihood approach, i.e. livelihood assets, activities, strategies, outcomes, institutional environment and the vulnerability context, are placed in this context. Vulnerability is related to the effects of HIV/AIDS on households' livelihood assets. The concept of household as the unit of analysis is discussed, as well as the gendered impacts of HIV/AIDS and their implication for agriculture. At the end of the chapter a conceptual framework is presented that depicts how these concepts are interrelated.

2.1 Theoretical approaches to agricultural technology development and adoption

2.1.1 Innovation-diffusion model

This model is composed of four basic theoretical approaches each focusing on a different element of the innovation process. These are combined to create a meta-theory of diffusion consisting of four components: the innovation decision process, the perceived attributes of the technology, the rate of adoption and individual innovativeness (Rogers, 1995).

Innovation decision process

The innovation decision process is characterised by five stages: knowledge, persuasion, decision, implementation and confirmation. In the knowledge stage the individual or household is exposed to the innovation's existence and gains understanding of how it functions. However, even after knowing about an innovation individuals may need to be persuaded to use it because they do not regard it as relevant to their situation. The outcome of the persuasion stage is either adoption or rejection of the innovation. The implementation stage is when an individual puts an innovation into use. The final stage is confirmation during which the individual seeks reinforcement for the decision made.

Attributes of innovations and their rate of adoption

Rogers (1995) identifies five attributes upon which an innovation is judged. These are relative advantage, compatibility, complexity, triability and observability. Relative advantage refers to the degree to which an innovation is perceived as better than the practice it replaces. Relative advantage is often expressed in terms of economic, social or other benefits. Compatibility refers to the degree to which an innovation is perceived by potential adopters to be consistent with their existing values or practices. Compatibility with what is already in place makes the new practice seem less uncertain, more familiar and easier to adopt. Complexity refers to the degree to which an innovation is considered as difficult to understand and use. If potential adopters perceive an innovation as complex, its adoption rate is low. Triability refers to the extent to which an innovation may be subjected to limited experimentation. Finally, observability refers to the degree to which the results of an innovation are visible to others.

Individual innovativeness

This theory posits that innovations spread gradually over time and among people resulting in various adopter categories. The result is an adoption process that forms a normal S-shaped curve when plotted over time (Rogers, 1995). Rogers attributes this distribution of adoption to the role of information, which reduces uncertainty in the diffusion process. Based on this argument Rogers has classified adopters into five categories: innovators, early adopters, early majority, late majority and laggards. Innovators are described as individuals who are venturesome, eager to try new ideas and willing to take risks. Early adopters are described as the local opinion leaders in the system who function as the role models and are quick to see the value of innovations. The early majority is formed by the largest category. These people only make a decision after they are convinced of the benefits. The late majority are cautious and sceptical persons who do not adopt until the large majority has done so. They are usually the relatively poor and are averse to risk. The last group of adopters is the laggards. They are suspicious of innovations and change agents. They are usually poor and seldom take risks.

The innovation diffusion model has several limitations. One of the major shortcomings of the model is that it generally assumes that the most important variable is information and the willingness of the individual to change. An individual is characterised according to his behaviour without considering factors that influence his behaviour. In reality many other factors are known to influence the adoption of an agricultural innovation. These include the farmer's objectives, the level of the resource endowments of the individuals, access to resources, availability of support systems and the characteristics of the innovation. For example, access to resources such as labour and land can limit the adoption of an innovation to a small number of individuals in a society. This could apply to HIV/AIDS- affected households whose labour and other productive resources are limited. Access to productive resources is also gendered, with women having less access than men. In such cases an innovative individual may be labelled as a laggard, while late or non-adoption is caused by lack of resources. Information and support services from the extension systems may also limit the spread of innovation by targeting innovators and early adopters while ignoring the others.

2.1.2 Economic constraint model

Economic theory perceives farm households as decision makers whose concern is how much to devote to the cultivation of each crop, whether or not to use purchased inputs, which crops to grow on which fields, and so on. This they do depending on their goals or objectives and the resource constraints of the individual farming household. The economic constraint model makes various assumptions. The model assumes that the household acts as a unified unit of production and consumption that aims to maximise utility subject to its production function, income and total time constraint. Utility is described as the satisfaction an individual derives from a set of commodities (including leisure), which is attained from consumption. A central feature of the model is the use of a single decision maker and the implicit assumption is that no conflict exists within the household and that all members have the same utility function so that maximising the household utility would yield similar results as maximising individual functions. This proposition is based on the assumption that household members will sacrifice their individual preferences for the common good of the household. The usual practice is to adopt the utility function of the household head, usually the man, to represent the utility of the entire household. In return, the altruistic head will make decisions based on what is best for the household as a whole.

In practice there are different tastes and preferences among the household members that cannot be left to altruistic behaviour. The assumption of altruism in the household is in contrast with the presumed selfishness and competitiveness of individuals. The single-household utility also obscures the likelihood of conflict and inequality in household decision-making. The popular idea of comparative advantage used in this approach (Evans, 1989), where individual household members are thought to specialise in those tasks at which they are relatively more efficient compared to other members, ignores the social relations that prevail in households (Ellis, 1993). These include non-market reasons for the division of resources and unequal power relations in decision-making. Empirical evidence has shown that different adult members of the household may conduct separate enterprises, earn individual incomes and manage separate budgets (Koopman, 1991). Many studies show that personality and individual circumstances rule out uniform predictable responses and that gender relations shape labour obligations and resource distribution (Kabeer, 1991; Whitehead, 1985). Households are now recognised as internally differentiated units usually characterised by cooperation and conflict on the grounds of gender and age (Sen, 1990). There is also the influence of custom and culture that are important in household decision-making (Pennartz & Niehof, 1999). Research has also challenged the conventional notion that ascribes household headship automatically to men. Women are now increasingly recognised as legitimate household heads in their own right (Mencher & Okongwu, 1993).

2.1.3 Central source of innovation model

Another dominant paradigm that has been used for a long time to explain agricultural research and technology diffusion is the central source of innovation model. In this model, also known as the transfer of technology (TOT) approach, innovations are seen to move progressively from the international agricultural research institutions, national agricultural systems, to national extension systems and finally to farmers (Biggs, 1990). The major emphasis in this model is on the transfer of knowledge and technology from research institutions to farmers. The key features of the model include assignment of clear-cut roles to specific institutions and groups of people. Research institutions have either an international or national mandate to conduct research, extension agents are only supposed to pass on the results, whereas farmers are seen as technology adopters or people who have problems that are fed back to extension advisers and researchers. The process of technology generation and transfer is seen as a linear process where scientists develop technology, demonstrate it to farmers through the extension agents, and the farmers adopt it in the final stages. In this model research institutions are the sole source of technology. The farmers' experience, knowledge and resources are overlooked and farmers are thus seen as passive receivers of technology (Leeuwis & Van den Ban, 2004). However, this paradigm has proved inadequate for managing the emerging challenges in agricultural research and technology diffusion today. These challenges include: (1) diverse biophysical environments, (2) multiple livelihood goals, (3) rapid changes in local and global economies, (4) an expanded range of stakeholders over agriculture and natural resources, (5) drastic decline in resource investment for the formal research and development sector, and (6) the impacts of HIV/AIDS on agricultural production (Gonsalves *et al.*, 2005). These new challenges suggest that research and development can no longer be the exclusive domain of scientists, but must be a joint process requiring the participation of a wider range of actors.

2.1.4 Multiple source of innovation model

This model, which is an improvement of the central source model posits to understand the clients' diverse needs and resources and views the users not merely as adopters but as active participants in the process of technology development and adoption. This model emphasises that agricultural innovations are derived not only from agricultural research institutions but from multiple sources. These sources include farmers, innovative research practitioners, research-minded administrators NGOs, private corporations and extension agents (Biggse, 1990). In the multiple source model, perspectives of the users of technology are seen as important in helping to develop and transfer locally usable innovations (Hardon-Baars, 1997). Furthermore, it redefines the role of farmers from being simply recipients to actors, who influence and provide inputs to the process. The needs of the users, who include women as well as men in farming households, are taken into account. In addition, the approach looks at the availability of institutional support required for successive adoption. The multiple source of innovation model encompasses the use of participatory approaches that have evolved from efforts to improve technology development and dissemination. Participatory methodologies are often characterized as being reflexive, flexible and interactive, in contrast with the rigid linear central source model. Experience has shown that innovations for improving agriculture need to address not only the technological but also the socio-cultural, political, economic dimensions such as community structures, gender, collective action, property rights, land tenure, power relations, policy and governance. This is especially more so in this era of HIV/AIDS and its projected impacts on agriculture.

One of the characteristics of participatory approaches lies in innovative adaptations of methods drawn from conventional research and their use in new contexts, in new ways, often by as well as with local people. The sustainable livelihood framework is one approach that has gained popularity as it employs participatory approaches in analysing interventions in people's livelihoods.

2.2 Sustainable livelihood approach

This study adapts the livelihood approach as the point of departure for analysing technology adoption. The approach draws on a number of theoretical and conceptual approaches that enable the identification of issues that are relevant in explaining livelihood strategies. These include issues that are far out of the domain of conventional research, e.g. issues related to HIV/AIDS-induced vulnerabilities, but may still be crucial to understanding important constraints on the ability of a technological intervention to affect livelihoods (Adato & Meinzen-Dick, 2002). The approach gives an improved understanding of both households' livelihood processes and how these are affected by interventions. Agricultural technology interventions often impact the livelihood activities, strategies and outcomes of households.

There are several definitions of livelihood all of which emphasize human agency and capabilities of people to shape their lives using material and non-material assets at their disposal to make a living. Ellis (2000:27) defines livelihood as 'the capabilities, assets (including both material and social resources) and activities required for a means of living'. Chambers & Conway (1992:7) define livelihood as comprising of 'the capabilities, assets (stores, resources, claims and access) and activities required for a means of living'. They call a livelihood sustainable when it 'can cope with and recover from stress and shocks, maintain or enhance its capabilities and assets, and provide sustainable livelihood opportunities for the next generation; and which contributes net benefits to other livelihoods at the local and global levels and in the short and long-term'.

Livelihood framework analysis is based on the assumption that people pursue a range of livelihood outcomes (health, income, reduced vulnerability, food security, etc.) by drawing on a range of assets to pursue a variety of activities. Their objectives drive the activities they adopt and the way they reinvest in asset building. Ellis (2000) perceives the livelihood approach as comprising three main dimensions: assets, the mediating processes that influence access to those assets, and the strategies adopted for survival. The livelihood approach therefore provides an approach that allows us to focus on the people, the assets they have, activities they engage in, and the outcomes they aspire to achieve, when studying the effects of the introduction of an agricultural intervention on people's livelihoods.

2.2.1 The household as the level of analysis

Despite the shortcomings of the economic household model highlighted above, the household is still regarded as the micro-level basic unit of analysis in this study. This is because households are seen as the basic institutions within which resource sharing and exchange among individuals takes place (Bruce & Lloyd, 1995). There are several definitions of households offered by different authors. According to (Ellis, 2000:18) 'Household is conventionally considered as the social group which resides in the same place, shares the same meals, and makes joint or co-ordinated decisions over resource allocation and income pooling'. Pennartz and Niehof (1999:3) define households as social units that enable individuals of different ages and sexes, to pool income from multiple sources in order to ensure their individual and collective livelihood. In this definition members of a household are not necessarily biologically related and may include friends, lodgers, servants who may not always share a common residence. Rudie (1995:228) defines the household 'as a co-residential unit, usually family-based in some way, which takes care of resource management and primary needs of its members'. This definition takes into account the family relation angle as well as the management of resources. Definition of the term household in the African context, where members of individual residential units are often embedded within strong networks of wider family and kin, may differ from the Western context. It may also be difficult to delineate household boundaries in such situations, and various studies have shown that household boundaries vary globally in relation to social and economic differences (Rudie, 1995). Kabeer (1991:9) concludes that the notion of the 'household is an analytical construct and its boundaries may have to be drawn pragmatically, in response to the concrete research questions being investigated'.

In this study, Rudie's definition that focuses on the family relation as well as the management of resources is adapted. According to Rudie, household definitions that solely focus on the family dimension have led to the stereotype tendency of looking at the household as a universal natural unit in all societies where women give primary care and are associated with nature. She argues that this focus is mainly based on Western ideals and tends to overlook the wide variation that is found in division of tasks between men and women, in kinship systems and residential arrangements. She criticizes an exclusive focus on resource management because this depicts households only as units organising production and consumption without taking into account the differences in societies, especially what she calls 'pre-capitalist societies of a peasant type' where both activities take place. She argues that there is need to take into account both dimensions, which help to explain the complex intra-household dynamics that occur in non-western households.

The study specifically focuses on farming households that are described as family-based, and who produce for both home consumption and market sale. The farming household is embedded in a system consisting of the farm production and the household unit where the former is a production unit while the latter ensures maintenance and reproduction. Depending on the resources and assets available to the household it engages in activities directed at satisfying the material needs of the household members. The collection of activities carried out to achieve a certain livelihood is called the livelihood portfolio. The ultimate aim of these strategies is achieving livelihood security (Du Guerny's, 2002; Niehof & Price, 2001).

2.2.2 Household livelihood resources and assets

The livelihood approach is based on the premise that the resource and asset status of people is fundamental to understanding the options open to them, the strategies they adopt to attain livelihoods, the outcomes they aspire to and the vulnerability context under which they operate (Ellis, 2000). Niehof (2004: 324) distinguishes assets from resources and says that 'resources are what you use, assets are what you have'. Swift (1989:11) on the other hand divides assets into 'tangible and intangible stores of value or claims to assistance'. He identifies three categories: investments, stores, and claims. The investment category includes human capital (education and health), individual productive assets (animals, farm equipment, houses and domestic equipment) and natural resources. Stores include food stores, items of value such as jewellery, and money in the bank. Claims include reciprocal claims from other households, claims on patrons (big men), on governments, and the international community. Finally he includes a category, which he calls collective resources that include soil conservation, water harvesting structures, irrigation systems and access to common property resources. Niehof and Price (2001) have grouped resources and assets according to level: personal, household and environment, making a distinction between tangible and intangible resources. At the personal level, physical strength, health and talents are grouped as tangible assets while skills, education, gender, experience and capabilities are intangible assets. At the household level, space, income, tools, buildings and livestock are the tangible assets, while experience, knowledge and information are grouped as the intangible resources. At the environmental level there are both natural and man-made assets. Among the natural assets land, water, soil and biodiversity are grouped as tangible assets. Infrastructure, biodiversity are grouped as man-made tangible assets while market, church, social/political institutions, support networks are intangible assets. All these lists contain similarities that can be summarized by the five capitals used in the DFID livelihood framework (DFID, 2001): human, natural, physical, financial and social capitals. In this study, the five capitals as defined below will be used.

Human capital

Human capital is considered as one the basic building block or means of achieving livelihood outcomes. (Ellis, 2000). Human capital represents the skills, knowledge, ability to labour and good health that together enable people to pursue different livelihood strategies and achieve their livelihood objectives. At the farm household level, human capital is a factor of the amount and quality of labour available; this varies according to household size, education level, skill levels, and health status.

Natural capital

Natural capital comprises land, water, rivers, forests and other biological resources that people utilise to generate a livelihood. One of the principal physical natural resource for a farmer's livelihood is land and any decision related to land use would be affected by his access to land and security of tenure. 'A secure place to live, free from threat of eviction, with access to productive land and natural resources are essential for rural livelihoods in sub-Saharan Africa' (Drimie, 2002:6). Any decision on technology adoption will be influenced by the farmer's access to land and security of tenure.

Financial capital

Financial capital has been used to refer to stocks of money that the household has access to (Ellis, 2000). These include savings held in the bank, credit, stocks and fungible assets such as livestock, jewellery, gold or food stocks. At the farm household level financial capital refers to savings held in the bank, access to credit in the form of loans, and stocks. Financial capital only becomes a useful asset in the household when it is converted into other forms of capital or is used directly for the purchase of food. This ease of switching between uses makes financial capital one of the fundamental assets.

Physical capital

This refers to basic infrastructure such as transport, shelter, water, energy, communications and production equipment, which enable people to pursue their livelihoods. Physical capital is considered a producer good as opposed to a consumer good, because it comprises capital that is used in economic production in order to create a flow of outputs that can be used in the future, while consumer goods are purchased or produced for direct use on material standards of living (Ellis, 2000). Physical capital often takes the form of man-made goods such as roads, irrigation canals electricity and piped water. Rural infrastructures such as roads are particularly important in relation to the utilisation of agricultural technology as they allow the transfer of information and link the farmer to the market. At the household level, farm equipment, housing, livestock (used for animal traction) and other personal household properties that can be converted into cash are considered as physical capital.

Social capital

Various definitions of social capital have been offered. According to the World Bank (2001) social capital refers to features of social organisation, such as networks, norms and trust that facilitate co-ordination and co-operation for mutual benefit. Narayan (1997:50) defines social capital as 'the rules, norms, obligations, reciprocity and trust embedded in social relations, social structures, and society's institutional arrangements, which enable its members to achieve their individual and community objectives'. Another popular definition of social capital emphasises the reprocity within communities and between households based on trust originating from social ties (Moser, 1998:4). Portes (1998:6) emphasises social capital's function in different contexts by defining it as 'the ability of actors to secure benefits by virtue of membership in social networks or other social structures'. Portes sees social capital as an asset that one can claim in times of need. At the household level, kinship networks, friends, neighbours and the community are important social assets. According to Putnam (1993) when people engage in networks and forms of organisations, they develop a framework of common values and beliefs that can become a moral resource or the 'glue that binds a community together' (Rankin,

2002:3). Networks can be both formal and informal and the latter are especially common for rural women improving household livelihoods. In this study social capital is represented by household membership in community organizations. The dimensions of this proxy will include group functioning, financial contributions to groups, participation in decision-making, and heterogeneity of membership and temporal changes in group participation.

This definition of livelihood capitals is used in this study while appreciating the debate surrounding the use of the word capital (Toner, 2003). Capital is conventionally seen as the stock of productive resources built up by human action which may depreciate, be consumed, utilized in production or be sold off. Under such a definition, it can be argued that natural and social 'capital' are not always appropriately termed 'capital' resources. In adapting the framework in this study it is appreciated that trade-offs may be quite difficult to analyse as certain types of capital are 'things' (such as land resources), while others are assets, which are accessed through particular processes (such as social claims). The capitals are thus used as a tool for more qualitative discussion of key issues, rather than a precise measurement framework.

2.2.3 Household livelihood strategies and activities

Depending on the assets people have, they engage in livelihood activities and develop strategies that best provide them with the desired livelihood outcomes. Deciding on what assets to utilize, when and how, constitutes a household's livelihood strategy. Different authors identify various types of strategies. Scoones (1998) categorises strategies into agricultural intensification and extensification; livelihood diversification that includes both paid employment and rural enterprises; and migration for income generation and remittances. Carney (1998) lists livelihood categories as natural- resource based, non-natural resource based, and migration. Ellis (2000) categorises them into natural resource-based strategies and non-natural resource based strategies (including remittances and other transfers). Various livelihood activities are carried out in the different livelihood strategies. These are conventionally divided into production activities and reproduction activities. Production activities are those that produce goods and services that contribute to income, such as growing crops, raising livestock, earning wages, making things, trade, providing services, and various activities that members of the household engage in. In some cases, there are one or two dominant activities, such as farming, fishing, or craftwork, but many households are involved in multiple production activities without one dominating activity. Reproduction activities, sometimes called household maintenance activities, are those activities that are not tradable but are nevertheless essential for the well-being of household members and the reproduction of the conditions through which a family survives. They include activities such as childcare, cooking, cleaning, caring for the sick, fetching water and collecting firewood. In this study, three categories of livelihood strategies i.e. agricultural intensification, livelihood diversification and migration are distinguished (cf. Scoones, 1998).

Agricultural intensification

Agricultural intensification is defined as 'increased average inputs of labour or capital on a smallholding, either cultivated land alone, or on cultivated and grazing land, for the purpose of increasing the value of output per hectare' (Tiffen *et al.*, 1994:29). Agricultural intensification may occur as a result of an increased (per fixed unit of land) frequency of cultivation, an increase in labour inputs, or a change in technologies (Carswell, 1997). Two factors may necessitate agricultural intensification. One is an increased demand for output as a consequence of population growth, in-migration or increased market demand in a country or region, or demand for higher value-added output (such as fruit, vegetables etc.) when income per head grows. Secondly, a decline in land/labour ratio may also necessitate intensification. Agricultural intensification is associated with a change in technologies and an increase in frequency of cultivation or labour inputs. For the process of agricultural intensification to take place effectively the institutional frameworks must be in place. These include a land tenure system that provides security in decisions to invest in land, credit institutions that provide capital for the purchase of inputs, and a functioning marketing system for the products of intensification.

Livelihood diversification

There is an increasing recognition that most families base their livelihoods around a wide range of activities that seek to maximize the use of resources and assets accessible to them. The literature on livelihood diversification is characterized by many terms and definitions. This study adopts the definition given by Ellis (2000:15), who sees diversification as 'the process by which rural households construct an increasingly diverse portfolio of activities and assets in order to survive and to improve their standard of living'. Livelihood diversification activities are generally classified on the basis of their roles as means for coping, adaptation and accumulation. Differences have been observed between poor households that are struggling to survive and better-off households that are diversifying to accumulate. Thus causes and motivations of diversification vary across families and for the same family with time (Niehof, 2004; Ellis, 1998).

Rural farm households diversify into on-farm, off-farm, and non-agricultural activities. On-farm diversification involves production of more than one crop, or production of different varieties of the same crop. Off-farm activities mostly include informal employment in agricultural activities in the local area or outside the area. Non-agricultural activities on the other hand are defined here as any work that does not directly involve plant or animal production. They include participation in trade, service provisioning, craftwork, or transfer payment in the form of state pension. A study conducted in Africa by the DARE program (De-Agrarianisation and Rural Employment) reports that stringent economic measures undertaken during the SAP implementation have contributed to a surge in non-agriculture income sources over the past 15 years. This is because as subsidies were removed from agriculture, education and health, the daily cash requirements increased while returns from farming became less. It is estimated that about 40 percent of African rural household income on average is derived from non-farm sources (Reardon, 1997; Ellis, 1998). Remarkably high levels of 55–80 percent were reported in DARE survey results Bryceson (1999).

Migration

Migration is a type of household livelihood diversification. It involves the movement of one or more members of the household for a certain period of time in search of a better and more secure livelihood for the household. Migration occurs due to what the literature terms 'push' and 'pull' factors. Income differences between the sending and receiving areas are one of the major pull factors, while land fragmentation (at inheritance), landlessness and seasonality are main push factors. Kenya has a long history of labour migration as a common livelihood strategy among the rural farming households. The culture of migration in Kenya is attributed to both push and pull factors. The urban centres created by the colonial administrators such as Nairobi, Mombasa, Nakuru, Eldoret, Kisumu, Kitale, Thika, Nyeri, Nanyuki, all located by the railway line, acted as pull factors for migration in Kenya. Subdivision of land and landlessness forced people to move out to urban and other areas. Indeed, migration became an integral part of most rural households. Berry (1993) reports that as early as the 1970s, rural households in Central Kenya were estimated to derive one-half of their incomes from non-farm sources. A large proportion of these incomes came from seasonal and longer-term labour migrants. Ellis (2000:70) distinguishes four types of migration:

Seasonal migration
This refers to temporary migration of household members during the slack labour periods and the return for the peak periods in the agricultural calendar.

Circular migration
This is movement of household members in response to cyclical needs for labour in non-farm labour markets. Migrants regularly return to their principal resident household as they do not set up alternative homes in the places they temporary go to work.

Permanent migration (rural-urban)
This involves movement of household members to a different location, mostly an urban area where they set up a residence. They regularly or intermittently send remittances back to their original home. This pattern, which was most common in Central Kenya, started during the colonial era and became more pronounced during the post-independence period. Overpopulation in the province had led to landlessness that was aggravated after independence in 1963 when most of those imprisoned or detained during the Emergency Period were released and came back home. Many of them found that their land had been taken over by the so-called 'loyal Africans' who had sided with the colonial government (Clough, 1998). Some were forced to migrate to the urban centres and particularly to Nairobi, which is not far from any part of Central Province.

International migration
In this pattern the migrant moves temporarily or permanently abroad. Limited international migration had been going on in Kenya for many years but increased from the mid-1980s due to lack of employment and a declining economy. There are now many Kenyans who have migrated and settled in foreign countries in search of greener pastures and promising labour markets.

Urban-rural migration.
A fifth category not usually included in migration literature is urban-rural migration. This is becoming a common pattern in Kenya, where people mostly in their retirement ages (55 years and above) are returning to their rural homes. Some of those returning have maintained links with rural areas and have built a permanent house where their wives and children remain. Some have at times invested in the rural economies. A new category of returnees is AIDS patients who go back home to be cared for by their relatives. Those returning under such conditions do not contribute much to the rural economy.

Remittances are amongst the most important aspects of migration. Remittances are invested in human or physical capital to enhance household production. A review of migration literature by De Haan (1999) suggests that although migration does not lead to far-reaching changes in agriculture, it plays an essential role in the maintenance of farming households' livelihood. It does provide for basic necessities, some consumerism, building up social capital and purchase of agricultural inputs.

2.2.4 Livelihood outcomes

Livelihood outcomes are the achievements and benefits that households anticipate to obtain through the implementation of specific activities and strategies. These outcomes can also be interpreted as the aspirations of the household. Potential outcomes include conventional indicators such as more income, improved food security, reduced vulnerability and more sustainable use of the natural resources (DFID, 2001). Outcomes may also include strengthened asset base and improvements in well-being aspects such as health. Outcomes are not necessarily the end point, as they feed back into the future asset base and the vulnerability status of the household. According to Niehof (2004:325), 'livelihood generation proceeds in a cyclical mode, which may take the form of either an upward or a downward spiral'.

2.2.5 The institutional environment of rural households

So far we have focused on livelihood resources and assets, whose combined use allow households to pursue various strategies and realize different outcomes. But resources and assets are only important to livelihoods if they can be accessed and support livelihoods in a sustainable manner. For the household there are endogenous and exogenous factors that influence access to assets and their use in the pursuit of viable livelihoods. Endogenous factors include social relations of norms and structures, which are part of the household. Intra-household gender relations that influence access to assets may affect the adoption and suitability of agricultural technologies. Exogenous factors consist of economic trends, policies, institutions, organisations and shocks. Institutions may be both formal and informal, ranging from tenure regimes to labour sharing systems, to market networks or credit arrangements. An understanding of institutions and organizations is important as they mediate access to livelihood resources and assets and in turn affect the composition of livelihood portfolios and strategies. Institutions that are considered in this study include agricultural extension, credit institutions, and marketing.

Limitations of the sustainable livelihoods framework

Despite the versatile nature of the livelihood framework it has been criticised for not capturing certain aspects that are important in explaining people's decisions and consequent livelihood outcomes. These include the aspect of culture, the notions of power and power relations and historical factors (Adato & Meinzen-Dick, 2002). Culture refers to beliefs, traditions, status, and identity (Brons et. al forthcoming). Although culture may not have direct economic value it is centrally important in people's lives, choices, and well-being. Another aspect that is not covered by the framework, which is important in shaping people's livelihood options and strategies, including their technology choices, is that of power relationships. Intra-household power relations are known to influence access to resources and consequently livelihood activities and outcomes. In this study these limitations will be incorporated by using the framework in conjunction with concepts drawn from gender theory and by taking the cultural context into account.

2.2.6 The vulnerability context in a situation of HIV/AIDS

Several definitions of vulnerability in livelihood studies have been offered (Chambers, 1990; Swift, 1989). Blaikie *et al.* (1994:9) defines vulnerability as 'the characteristics of a person or group in terms of their capacity to anticipate, cope with, resist and recover from the impacts of natural hazard'. Chambers (1990:2) distinguishes vulnerability from poverty by stating: 'vulnerability is not the same as poverty. It means not lack or want, but defencelessness, insecurity, and exposure to risk, shocks, and stresses'. A common theme running through much of the literature on vulnerability is a link to coping and the capacity to handle stress or shock. Watts and Bohle (1993:118) distinguish vulnerability into three dimensions: the risk of exposure to crises, stress and shocks; the risk of inadequate capacities to cope with stress crises and shocks; and the risk of severe consequences of and the attendant risks of slow or limited recovery (resilience) from crises, risk and shocks.

Vulnerability is also distinguished by its external or internal dimensions (Chambers, 1990; Brons *et al.* forthcoming). Internal vulnerability refers to the specific characteristics of an individual or household (age, sex, education and health status), which make them unable to cope with stress, crises or shocks. Internal vulnerability is manly associated with lack of access to assets. This is because one of the means of resistance to vulnerability is the assets and entitlements that individuals or households can mobilize and manage in the face of hardship. According to Moser (1998:3) 'the more assets people have, the less vulnerable they are, and the greater the erosion of people's assets, the greater their vulnerability. External vulnerability refers to risks of exposure to crises, stress and shocks that are beyond an individual's control. These include changes in climate, political changes, war and terms of trade'.

The impacts of HIV/AIDS form both the external and internal vulnerability context of households. The pandemic contributes directly to internal vulnerability through its impact on human capital such as changes in household age structure, gender of headship and loss of labour and skill. Indirectly, HIV/AIDS adds to household vulnerability through the sale and disposal of assets to meet medical costs (Barnett & Whiteside, 2002). The erosion of households' assets also contributes to internal vulnerability. Literature has shown that in those parts of sub-Saharan Africa where the epidemic has reached peaks, rural households are facing health care costs they cannot meet without selling assets or going into debt. As the cost of death and illnesses rise, there is reduction or liquidation of any savings they may have as well as a reduction in their asset base of equipment and animals. Apart from material loss rural households are also losing social solidarity through loosening of social bonds and the thinning of social fabrics (Rugalema, 1999). What used to be safety nets for most rural folks have now become 'safety nets with holes' that can no longer be relied upon for support in times of crisis (Baylies, 2002). Müller (2004, 2005a, 2005b) presents an overview of the impacts of HIV/AIDS on livelihoods of rural farming households in sub-Saharan Africa. In the next section the effects of HIV/AIDS on the five livelihood capitals and the consequent implications to agriculture and technology adoption are briefly discussed.

Effects on human capital

When HIV/AIDS strikes it is the domestic-farm-livelihood-labour interface, which experiences the stress of the impact, particularly in small-scale African agriculture that relies almost exclusively on family labour. By attacking the able-bodied and active adolescents and adults, HIV/AIDS undermines the farm-household through the direct loss of labour for the farm and of time available for both farm and household tasks. Gillespie *et al.* (2001) proposes that the impacts of AIDS will be felt along different time scales. Morbidity due to frequent illness from opportunistic diseases such as tuberculosis reduces labour productivity. This also has indirect effects as some of the adult household labour is withdrawn or diverted to caring for the sick and attending funerals. This has several implications as discussed by Haddad and Gillespie (2001). The initial response is what they call the 'importation' of labour. This involves bringing in relatives or other members of the household previously living elsewhere. Poor households may not have the cash to hire labour and even where the household has resources to hire labour this may not be available due to labour migration. The second option is reallocation of the remaining household labour. This means that the remaining members of the household undertake extra activities and if they are women this means an addition to their working hours. Children are also withdrawn from school to help in the household, girls being the first to be withdrawn.

As well as HIV/AIDS impact on reduction of labour, there is also a loss of farm-specific knowledge as a result of changes in the age structure and quality of skilled and unskilled agricultural labour (Topouzis, 2000; Haddad and Gillespie, 2001; Müller, 2004). Due to mortality there is a growing number of elderly people and children who assume a greater role in farming. Premature death does not allow for transfer of farming knowledge to the younger and older generations that are faced with the challenge of farming in changing agricultural environments. In Africa, it is now acknowledged that even so-called unskilled labour requires task-specific skills that are very difficult to replace (Cohen, 2002). This study assesses to what extent family labour in banana farming households has been influenced by the epidemic and how this influences their ability to utilize the tissue culture banana technology.

Effects on natural capital

A study conducted in three African countries on the impact of HIV/AIDS on land found that affected households took various options which included leaving land fallow, abandoning the land, selling the land, entering into sharecropping or changing land use as they get into less labour intensive crops (Drimie, 2002). The long-term effects on agriculture are predicted to be changes in farming systems as household cultivation shifts from cash crops to subsistence crops and from labour-intensive to labour-extensive crops that are often also less nutritious. The emphasis shifts to meeting food needs first, which leads to decrease in cash income at the household level (Gillespie, 1989). It has been shown, for instance, that in Bukoba District, Tanzania, a labour-extensive, low-external-input cassava/sweet potato farming system has replaced the intensely managed banana/coffee/bean farming system (Rugalema, 1999).

The effects of HIV/AIDS on land are also being felt through tenure rights of surviving widows on customary land (Drimie, 2002). Since ancestral land is still distinctly inherited through patrilineal patterns in Central Kenya, widows and orphans are increasingly losing inheritance rights in the HIV/AIDS era. A study by the Forest Action Network (FAN, 2002) in Bondo and Nyeri district in Kenya reported of cases of women being dispossessed of their right to use land after their husbands' death. Widows in cases where the deceased has died of HIV/AIDS are often condemned as the ones who have infected their husbands and are subsequently under massive pressure to leave their marital homes, thereby losing their rights to access and use of land. This has implication for land use in relation to technology adoption.

Effects on financial capital

Households affected by HIV/AIDS turn to a number of different coping strategies, most of which lead to a decrease in household financial capital. Households are forced to use their income, and savings for medical care, funeral expenses, and other immediate expenses. AIDS decreases agricultural production income by removing from the labour force not only the sick person, but also other members of the household who must care for the patient. According to report from the Joint United Nations Programme on HIV/AIDS (UNAIDS, 1999), families in Côte d'Ivoire, Tanzania, and Thailand who were coping with HIV/AIDS experienced a fall in income of 40 to 60 percent. Given that most technologies developed to increase agricultural production are based on the purchase of inputs such as fertilisers and seed, financial capital is a crucial resource. With the depletion of farm and off-farm income due to HIV/AIDS most households are unable to adopt technologies that are input-intensive and may turn to low-input production that reduces production both for home consumption and market. In cases where affected households have sold off their assets, they cannot access rural microfinance institutions for lack of collateral. Also, affected families increasingly face difficulties finding co-signers for taking out a loan.

Effects on physical capital

To raise cash to pay for health care or food and funeral expenses, families affected by HIV/AIDS sell food-producing assets, such as cattle, chickens or goats and farm equipment. Some of these assets play a complementary role in subsistence farming. For example, where a household keeps one or two cows under zero-grazing, a few chicken and goats, these provide farmyard manure that is used to improve soil fertility in case the household cannot afford fertilisers. The animals are in turn fed on the stalks and pseudostems (in the case of banana) produced from the farm. Thus, the sale of such assets besides reducing the household's food security also negatively affects soil fertility. A study carried in Namibia showed that in those households where a male household head had died 52 percent of them had lost cattle, 30 percent small animals and 39 percent farm equipment (Ekaas, 2003). The sale of productive equipment especially needed for agricultural production compromises the household's ability to generate income in the short term. Therefore it is necessary to look at how ownership of physical capital influences households' ability to adopt technologies.

Effects on social capital

Social capital may be undermined by HIV/AIDS in several ways. First, the incentives for coordinated group action may be diminished as a result of the heavy discounting of the future benefits of such action. One of the mutual benefits that is enhanced by social capital is accessibility to informal credit through group-based micro-finance programs. Availability of funds to carry out timely purchases of cash inputs into agricultural production, as well as to buy capital equipment like water pumps, has long been regarded as a critical constraint inhibiting productivity increase in small-scale agriculture. The severity of this constraint is mainly in the poor functioning of rural financial markets in developing countries in terms of the high costs of securing adequate information on potential borrowers, the risk of default on loans, and the lack of collateral to put up against loans. As noted earlier, HIV/AIDS-affected household are often unable to adopt technologies that require purchase of farm inputs as their financial capital base is depleted. Group-based lending, which is an alternative source of credit for these households, may now be challenged as affected members become a liability for the group. The micro-finance institutions involved in these group-based lending programs are also faced with the challenge of repayment and of affected clients. Second, HIV/AIDS-affected households may be excluded from such social networks due to stigmatisation. According to Narayan

(1999) the stigma attached to HIV/AIDS is not conducive to the establishment of ties across the different strands of social capital in communities and this might lead to exclusion. Thirdly, the formal institutions that also contribute to social capital formation, such as church groups, women groups, and youth groups are likely to be weakened. Membership loss due to death and lack of time of affected members to attend and be involved in group meetings may result in disintegration of such groups.

It is generally agreed that social networks that are more heterogeneous and spatially concentrated tend to be stronger and have higher benefits than homogenous networks (Gillespie *et al.*, 2001). However, this advantage may introduce a fourth undermining factor in that members who are highly mobile or live in urban areas will make a network more susceptible to HIV/AIDS. Lastly, social reproduction in terms of role-modelling deteriorates. Future generations not only do not learn these farming practices, they also do no experience the informal exchanges of knowledge that are often embodied in farming livelihood practices.

Impacts of HIV/AIDS on institutions

HIV/AIDS increasingly impacts on and changes the institutional environment. In the case of government departments or civil society institutions, people and clientele, as well as ways of working with people, will change. The internal capacity of organisations will also be affected as more staff becomes infected. Most notably, as infection rates increase, absenteeism increases which leads to decrease in staff productivity. This are coupled with increasing financial costs to the institution in retraining staff to replace those who fall ill and die. Staff turnover will increase as individuals get sick and need to be replaced.

Extension services

Although farmers may be experts in what they do, intensified production requires information and training on methods and the scientific properties of the inputs and their application (Doss, 2001). This information is traditionally provided by government extension services. The Structural Adjustment Programs (SAPs) have crippled many of these services. During the SAPs period, reforms were introduced in the civil service, which included major retrenchment programs that were accompanied by government budget rationalisation programs. In Kenya, the government reduced to half the expenditure in the agricultural sector, which had dire consequences for the extension services. The HIV/AIDS epidemic has aggravated the situation. A preliminary study conducted in Kenya in 1999 showed that 58 percent of all deaths in the preceding five years in the MOA had been AIDS-related (Guenter, 1999). Topouzis (2000) estimated that levels of HIV/AIDS prevalence among MOA staff were likely to be at least as high as national average estimates. The effects of morbidity and mortality in agricultural and related institutions have been analysed by several authors (Loevinsohn & Gillespie, 2003; Topouzis, 2000; Haddad and Gillespie, 2001). One effect is the increases in the workload of Ministry of Agriculture (MOA) staff that leads to severance of key linkages in the organisation and service delivery chain and compromises the quality, scope and continuity of agricultural extension services. The second effect is the impact that the disease has on depleting the pool of highly skilled and specialised personnel. There is also a drain of MOA resources due to the direct/indirect costs of HIV/AIDS such as medical bills, life insurance claims, death gratuities and funeral expenses that are likely to result in reduced funds for service provision. Even though MOA staff may be confronted with HIV concerns on a daily basis, they are constrained by lack of skills of how to respond to the epidemic, as it is not part of their formal education and training. Technology adoption heavily relies on the farmers' accessibility to good-quality and appropriate information. The effects of HIV/AIDS on extension services compromise all this.

2.3 The concept of gender

Gender refers to socially constructed maleness and femaleness and varies widely across cultures. The role of gender in livelihood generation cannot be over-emphasized. This is because men and women bear different responsibilities and have different options in household livelihood generation. The concept of gender is used to describe all the socially given attributes, roles, activities and responsibilities associated with being a male or a female in a given society.

2.3.1 Gender and agriculture

The central role of gender in agricultural development in African agriculture is now widely recognised. Women produce 80 percent of the basic foodstuffs both for consumption and sale in sub-Saharan Africa, yet their key role as food producers and providers and their critical contribution to household food security is only now becoming fully recognised (FAO, 2006). In Kenya, women supply 70-75 percent of agricultural labour and their contribution has been referred to as the backbone of the household production (Spring, 2000). While women are the main food producers, they lack access to and control over the means of production such as secure land tenure, information, credit and control of labour. When looking at gender-linked differences in accessing productive factors it must be acknowledged that gender roles and responsibilities are dynamic and change with new circumstances. It is therefore difficult to tell in advance what the effects of a particular intervention will be on a group of people (Doss, 2001; Ellis, 2000). Agricultural technology adoption and impacts depend on complex interactions and there is need to understand intra-household dynamics embedded in the complex and highly heterogeneous African farming households. Regardless of the targeted beneficiaries, individuals with greater power and access to resources are better able to take advantage of a change in circumstances than less powerful and poorer individuals.

Gender division of labour

New technologies are known to change labour allocations within farming households. Since Esther Boserup's (1970) study which highlighted the significance of women's work in agricultural systems, a plethora of empirical studies have been conducted to make the work of rural women quantitatively visible. The underlying ideology in these studies is that men were associated with the public workplace, while women were associated with the private or domestic realm (Spring, 2000). Embedded in this argument was that capitalist mode of production regarded only cash-cropping activities as work and women's domestic work of maintaining households, reproduction and food production disregarded. This led to the distinction of men's crops and women's crops (Koopman, 1993). Women were said to produce subsistence crops because they are responsible for feeding the family, while men grew cash crops because they were responsible for providing cash income for the family (Negash & Niehof, 2004). Throughout sub-Saharan Africa women tend to provide more labour for agriculture than men do (Agarwal, 1985; Carr, 1985.) The gendered division of labour that assigns to women the triple role of reproduction, production and community management leads to heavier workloads for women in comparison to men (Moser, 1989). Much of this work is unrecognised and unpaid, so that women are often over-employed in terms of hours worked and underpaid in terms of income received. Women's overall responsibilities affect poor households' capacity to adopt new activities when additional family labour is not available or they cannot afford hired labour. Thus, though improved agricultural technology may lead to increased productivity, women may fail to benefit from it. Moreover studies have shown that woman's labour burden increases with new technologies (Doss, 2001; Doss & Morris, 2001).

Access to productive resources

Historically, women's access to land in most African cultures was based on status within the family and involved right of use, not ownership (Whitehead, 1985). Under customary systems of land tenure property is held in a man's name and passed partrilineally within the group. Women have access to land through their husbands while daughters do not inherit land and divorced women loss their ex-husbands land (Baerends, 1994). In patrilineal systems, which prevail in Kenya, land is mostly owned and controlled by men with traditional tenure and inheritance based on patrilineal descent (Spring, 2000). Among the Kikuyu, women traditionally derived their land-use rights from their fathers until marriage and from their husbands after marriage. Women had the right to work on their husband's farms and control the use of foods crops they themselves produced. However, the introduction of the Swynnerton Plan of 1954, which was a colonial government program for the registration of private title deeds to agricultural land, resulted in diminished security of tenure for non-title holders, notably wives, who could no longer rely on established secondary rights or kinship ties to guarantee land access. Although married women still have user rights over their husbands' land, the husband has now gained more exclusive rights over the land's disposal as it is now registered in his name resulting in what has been termed 'rigiditization' (Aliber & Walker, 2006). This has been described as the situation where men have absolute ownership of land to the exclusion of women and children. Thus, the usufruct rights that wives had to land through marriage have been reduced. Continued use of land across seasons and investments on its improvement is therefore dependent on the good will of men. Unmarried or divorced women have very few options for gaining land in this system land and are often reduced to begging from male relatives for rights to land use (Spring, 2000).

Thus, developing technologies that require secure land tenure will disadvantage those with insecure tenure and in most instances these are women. Furthermore, it is not uncommon to find that interventions aimed at improving productivity on land worked by women may end up by the land being taken over by men when there is a prospective of greater cash-earning opportunities and prestige. Ellis (2000:157) concludes 'since land ownership is one of the most powerful means of raising funds for everything from education to physical capital and to starting an own business, women's exclusion on these counts can also quite often be traced to their lack of rights in land'.

Gender differences also occur with respect to access to financial capital in form of credit since women are unable to provide the collateral that is required for a loan. Group lending approaches that mainly target women has redressed this. However, gender relations may sabotage these lending schemes where the men utilise their wives loans for their own projects and the women just become guarantors (Ali, 2005).

Control of income

Introduction of new technology can change income distribution within households. Where men and women play different roles in producing agricultural outputs, a change in technology may affect these roles with possible implications for the control over income. Studies indicate that it is usually men who have primary control over the family's cash income in many developing countries (Agarwal, 1985). A study in

Central Province, Kenya reported that whilst women in the area have taken over many of the roles that men used to perform, 'men still mostly make the major decisions and control the purse strings' (Kiriti *et al.*, 2003:119).

2.3.2 Gender and HIV/AIDS

Gender inequality is acknowledged as an important determining factor on the impacts of HIV/AIDS epidemics. In the first place women are biologically more susceptible to HIV infection than men (De Bruyn, 1992). Secondly, social, economic and cultural inequalities (such as lack of employment opportunities, poor access to education and information, etc.) make women more vulnerable to HIV infection and AIDS impact than men. It is estimated that 60-80 percent of African women with HIV have had only one partner but were infected because they were not in a position to negotiate for safe sex or prevent their partners from having additional sexual contacts (Barnett & Whiteside, 2002). Furthermore women in rural areas, tend to be even more disadvantaged due to reduced access to productive resources such as land and credit. Given the prevailing gender division of labour, HIV/AIDS disproportionately adds to women's workload as, in addition to their productive work, women are also the predominant care givers for the sick (Müller, 2005a; Wiegers *et al.*, 2006).

Access to information and farmer's knowledge is correlated to gender and thus has an impact on technology adoption (Doss, 2001). Ventura-dias (1985) found that in Kenya even where women are the actual farmers, it was through the male presence that these women farmers had access to agricultural information services. In addition, most of the extension services are focused on cash crops rather than food and subsistence crops, which are the primary concern of women. Gender related inequalities in women's access to productive resources such as land; credit information and education have led to the failure of most women being able to realize their full economic potential and consequent independence. Thus women may continue living in situations in which they are at risk of contracting HIV because they are dependent on the man.

2.4 Conceptual framework

Following the above discussion, this study takes the sustainable livelihood approach to analyse the suitability of the tissue culture banana technology in HIV/AIDS-affected farming households. This approach is used because it looks at 'where people are, what they have, their needs and interests' (Chambers, 1983:1). As opposed to regarding the farmer as a passive recipient of technology, the livelihood framework recognises him as an actor with assets and capabilities, which enable him to pursue his livelihood goals. The conceptual framework in Figure 2.1 shows

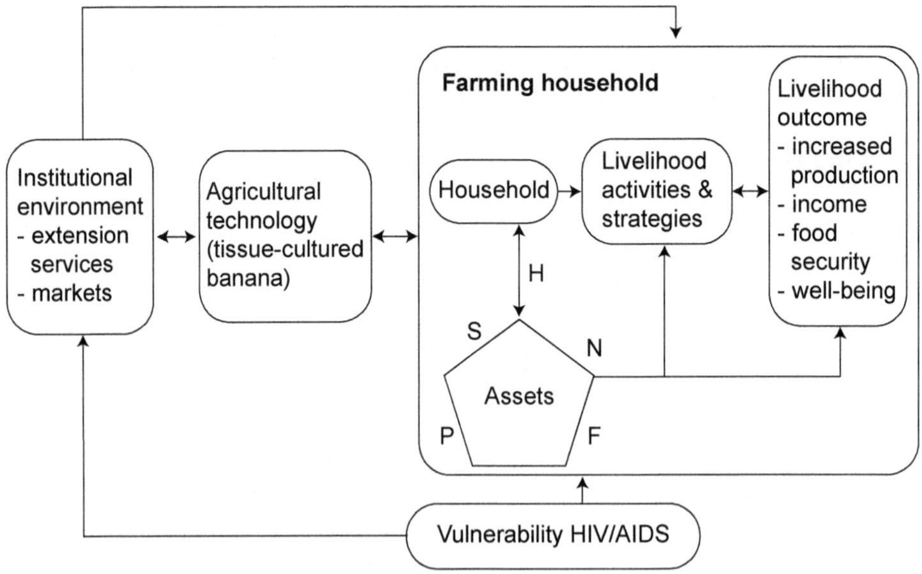

Figure 2.1. Conceptual framework.

the linkages between the various livelihood components, and how they relate to the desired outcome i.e. livelihood security. The box in the centre depicts the farming household consisting of the assets, farming household characteristics, livelihood activities, and the livelihood outcomes. Because asset endowments vary among farming households there is need to acknowledge this and recognise that their livelihood activities, strategies and outcomes differ. Response to shocks and interventions introduced in their livelihoods will differ depending on resource endowment. Notwithstanding these differences, farming households deploy their assets and resources to engage in multiple activities that ensure livelihood security subject to the strategic livelihood objectives. Farm families engage in farming both for own consumption and the market, and they often work in off-farm enterprises to earn cash for investing in future food production and other household needs. Many rural farming households benefit from remittances sent by relatives working in urban areas or abroad. This observation suggests that rather than focusing on food production *per se*, we should determine how small-scale agriculture activities support households' multiple livelihood strategies in an agricultural context. Livelihood outcomes include increased production, income, food security and well being of the household. Outcomes also feed back to the asset base and livelihood activities. The effects of HIV/AIDS on the farming household represent the vulnerability context, affecting the assets, activities and livelihood outcomes of the farming households. The first box represents the institutional environment that influences farming household's use of assets in pursuit of different livelihoods. This includes the extension services, land tenure policies and the markets. In adopting the livelihood approach it is acknowledged that gender differences exist among the individuals within farming household. Some individuals may have limited access to assets and are more constrained in their choice of livelihood strategies than others. Different individuals within a household may experience differing risk factors. These aspects are incorporated by integrating a gender perspective in the analysis.

Chapter 3

Study design and data collection

This chapter starts with a brief description of the study area followed by a detailed discussion of the research strategy and study design, the methods of data collection and analysis. The chapter concludes with a discussion of ethical considerations and of the problems encountered in data collection.

3.1 Research area

The study was conducted in the Republic of Kenya which lies across the equator in Eastern Africa, on the coast of the Indian Ocean. The country borders Somalia to the east, Ethiopia to the north, Tanzania to the south, Uganda to the west and Sudan to the northwest. It lies between Latitude 4° North to 4° South and Longitude 34° East to 41° East and occupies a surface area of 581, 751 km^2 of which 2.3 percent is water mass. Only 9 percent of the land is arable while 37 percent is pasture and the remaining 54 percent is semi-arid (FAO, 2004). Kenya has diverse physical features, which are divided into four distinct regions: the coastal belt, the low lying arid and semi-arid plateau, the highlands and the lake basin around Lake Victoria. The country is bisected by the Great Rift Valley. Kenya has the second highest snow-capped mountain (Mt. Kenya) in Africa after Mt. Kilimanjaro in Tanzania, rising to a height of 5,199 meters.

Kenya exhibits a wide range of climatic conditions. Except for the Coastal and Lake Victoria regions, altitude is the main determinant of rainfall with variations induced by topography. Generally the climate is warm and humid at the coast with rainfall levels of over 1,000 mm per year in April to July. The lake basin is also humid and wet where rainfall over 1,000 mm occurs reliably from March to November. The central highlands have a bimodal rainfall pattern, with the rainy seasons in March-May and October-December. Here precipitation is determined by the altitude with high-altitude areas (over 1,500 m) receiving substantial rainfall reaching over 2,000 mm per annum. However, here topography also has a major influence where even higher areas than 1,800 m that are in the rain-shadow areas of Mt. Kenya and the Aberdare Mountains are quite dry. The remaining part of the country falls into the semi-arid and arid zone with rainfall averaging less than 500 mm per annum with variation from year to year.

According to UNFPA (2005), Kenya has a population of 34.3 million people with an overall annual growth rate of 2.4 percent. With this growth rate the population is expected to rise to 83.1 million by 2050. More than half (61%) of the population is estimated to be living in the rural areas. There are over 70 ethnic groups categorized into: Kikuyu (22%), Luhya (14%), Luo (13%), Kalenjin (12%), Kamba (11%), Kisii (6%), Meru (6%), other African (15%), non-African (Asian, European, and Arab) (1%). These ethnic groups fall into three linguistic groups; Bantu, Nilotic and Cushitic with Kiswahili as the national language and English as the official language. The majority of the population is Christian (66%), although there is a proportion of traditionalists (22%), and Muslims (7%).

Maragua is located in Central Province of Kenya. The province borders Nairobi city in the south, Mt. Kenya in the north and the Aberdare ranges in the eastern side. It is divided into 7 districts; Kiambu, Thika, Maragua, Murang'a, Kirinyaga, Nyeri and Nyandarua. Maragua district which initially formed the larger Murang'a district lies between the altitude of 1,100 and 2,950 metres above sea level. The area receives a bi-modal type of rainfall with an average of 1,200 mm during the long rains (March to May) and 1,000 mm during the short rains (October to December). The district lies in nine agro-ecological zones ranging from the upper highlands on the slopes of the Aberdare range to the lower midland zones. Over 70 percent of the district lies in the Upper midlands and lower highland zones, and this is where most of the agricultural activities are carried out (Jaetzold & Schmidt, 1983). The study was conducted in Nginda and Ichagaki, which are located 1,340-1,670 m above sea level and are in the main coffee and marginal coffee zones (LM4 and LM5 agro-ecological zones) respectively.

This area was chosen as a study site because of various reasons. One is that banana, which for a long time was considered as a semi-subsistence women's crop, has become an important commercial crop in the area serving the urban markets of Nairobi and its environs. Previously small-scale farmers in the area depended heavily on proceedings from coffee for their livelihood but with the decline in coffee prices farmers had to look for other sources of income by diversification and commercialisation of traditional food crops such as the banana. The banana has the potential for food and livelihood security as it can both be consumed at home and sold in the market for cash. Despite the potential in banana production diseases and pests have limited production in the area and to counter this problem KARI, in collaboration with other development partners, introduced the tissue-cultured banana technology. This was done by supplying tissue culture plantlets of new banana varieties that were tolerant to the prevalent disease and pest-free. However, in spite of the potential of the technology for reducing poverty in the area, various factors, amongst them the relatively high prevalence of HIV/AIDS (5.6%) may hamper adoption and utilisation of the technology. Thus, the area was therefore chosen as it provides opportunities to study the significance of agriculture technologies for vulnerable farming households. The map in Figure 3.1 shows the location of Maragua distict.

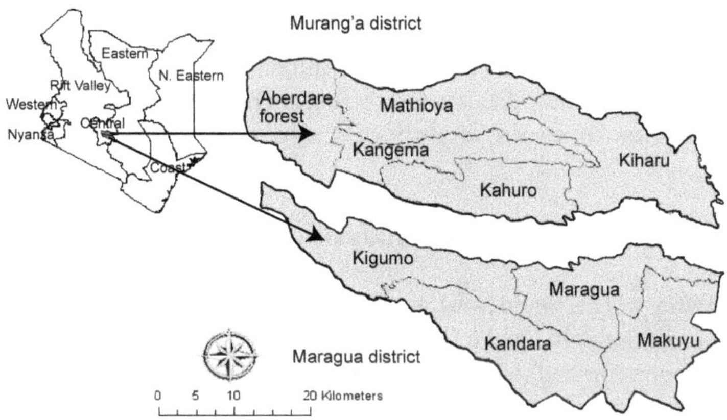

Figure 3.1. Map of Kenya showing the location of the greater Murang'a and Maragua District.

3.2 Research strategy

The study has an exploratory and descriptive design. A mixed-method approach combining both quantitative and qualitative methods was used. Due to the sensitive nature of collecting data in HIV/AIDS-affected households, it was difficult to use one single research method. Various scientists now recognize the importance of using both types of methods (Green *et al.*, 1989; Scrimshaw, 1990; Ritchie, 2003). Green *et al.* (1989) gave five ways of applying a mixed-methods research strategy: triangulation, complementarity, development, initiation, and expansion.

Triangulation is defined as the use of different methods of collecting data with the aim of comparing diverse aspects of the same phenomenon. It is often thought to help in validating or verifying the accuracy of information. Complementarity refers to adding information about results from one method with the results obtained by another. Development 'uses results from one method to help develop or inform the other method' (Greene *et al.*, 1989:259). Thus results from one method shape subsequent methods or steps in the research process. Initiation stimulates new research questions or challenges results obtained through one method. In the process new insights may emerge that were not planned in the research design.

This study used the triangulation and development strategies. Qualitative research methods were used at the beginning of the study to identify the appropriate dimensions of various concepts and also for generating the 'real life' language in which the questions for the survey would be framed. After the quantitative survey, in-depth interviews were used to contextualize, interpret and help understand the participants' perspectives. This involved a detailed study of sensitive and delicate issues such as gender relations in resource allocation and impacts of HIV/AIDS on households' livelihoods. Quantitative research on the other hand was used to answer questions that were mainly descriptive in nature that would enable generalisation.

3.2.1 The household as a unit of analysis

As explained above, the household was used as the basic unit of analysis in this study. However, at the field level it became apparent that the criteria for determining household membership had to be adapted. For example, male household heads who had migrated and lived in the city but always return home over the weekend or on a monthly basis or send back remittances were counted as household members. Conventionally, such members would not be included as they are not involved in the day-to-day decision-making. But in the research area members of the household still designated a husband who had migrated to and lived in the city as the household head. In this study a household member was defined as someone who had lived in the same house or in the same compound or homestead, shared food from a common source, and contributed to the resource pool of the household for a period. Additionally, husbands who worked in urban areas and came home at the end of the month or sent money every month were counted as household members as well, because despite their not being involved in the day-to-day decisions and activities in the household, they were very much involved in making major decisions such as sale or purchase of major assets, education of family members, and burial and marriage ceremonies.

3.3 Data collection

The data was collected in four stages: orientation phase; key informant interviews; focus group discussions; farm-household survey; and in-depth interviews. These stages are briefly described in the following section.

3.3.1 Orientation phase

This was carried out through a reconnaissance visit to the study area at the proposal writing stage. The objectives of the visit were to explore the proposed study area; to gain some general insights into the nature of the problem, and to establish contacts with key informants on the ground. It involved collection of secondary information on crop production and the HIV/AIDS situation in the area. During the visit discussions were held with the Ministry of Agriculture (MOA) extension agents on the trends in crop production over the last few years. The major banana producing areas of the district were identified to select the specific location for the study. Reports of Participatory Rural Appraisals (PRAs) that had been conducted in the district by MOA staff provided data on crop trends, seasonal calendars, historical profiles, maps, the daily division of labour between men and women, and the functioning of institutions (extension services, markets, etc.). The PRA reports had data on the main constraints identified in the area which were ranked in order of importance as: food insecurity; poor roads; human diseases; insecurity; soil infertility; unclean water; poor marketing structures; unemployment; and fuel wood shortage. Visits were made to several farmers in the two locations that were growing both tissue-cultured and non-tissue-cultured bananas and this helped in establishing contacts that proved to be very useful later on in the project implementation process. Information was also collected on the results of socio-economic studies that had been carried out in the area by KARI, highlighting the lessons learnt from the tissue-cultured banana project.

A visit was made to the Maragua district hospital, which has a voluntary counselling and testing centres (VCT) that collects sentinel surveillance data from the Antenatal Clinic (ANC). Antenatal clinics have been the principal tool for monitoring the spread of HIV in the country. The centre is a comprehensive care unit that coordinates testing for HIV/AIDS, counselling of AIDS patients, sensitization and awareness activities for HIV/AIDS in the community. It alo provides free anti-retroviral drugs (ARVs), with the support of the Kenyan and Spanish governments. Discussions were held with the staff at the centre focusing on the prevalence rates in the district over the years and the methods used in collecting the data at the clinic. During the discussions it emerged that HIV/AIDS prevalence rates in the district had gone down from 7.9 percent in 2002 to 5.6 percent in August 2003 as indicated by the ANC data. This was despite the fact that they were able to get a larger sample of the target groups that not only included expectant mothers but also commercial sex workers and people infected with sexually transmitted infections (STIs). This was mainly through campaigns by the team which had been successful in attracting more people to go for voluntary testing and counselling. The contacts established through this visit were crucial for identifying HIV/AIDS affected households later in the study.

3.3.2 Key informant interviews

The rationale for conducting in-depth discussions with key informants was that these are individuals who are grounded in the community and have particular or 'expert' knowledge about the area, its people, their livelihood activities, and HIV/AIDS prevalence and impacts in the area. Their insights were highly useful in understanding the nature of the problem. They also assisted in the selection of individuals to be involved in the various focus groups. Efforts were made to include key informants representing different demographic characteristics and value orientations. The key informants were drawn from eight categories of people with at least one person from each group. These groups were: agricultural extension providers; agricultural researchers; banana growers; social services workers; community leaders; women traders; people living with HIV/AIDS (PLWA); and administration officers (chiefs and sub-chiefs).

The interviews were conducted using guides which varied in content depending on the group the informant represented. However there were common questions for all informants such as their perception of the impact of HIV/AIDS on peoples' livelihood; gender division of labour and access to resources; and suitability of the tissue-cultured banana in a situation of HIV/AIDS.

3.3.3 Focus group discussions (FGD)

These were used as an exploratory tool to discover peoples' thoughts and feelings and to obtain detailed information about various subjects. Data were generated from the interaction between members of the group. The group setting was generally characterised by a synergistic effect whereby participants explored different points of view, and formulated their own ideas and understandings. This kind of exchange resulted in a lot of information being generated which was useful for identifying key issues; insights into needs, expectation, attitudes, perceptions and feelings; and developing interview schedules for the household survey questionnaire. Three key aspects were considered in conducting the focus group discussions. One was the selection of participants. Participants were purposefully selected on the basis of their experience related to the research topic. The second crucial factor was the composition of the focus group. The issues considered included whether people with similar characteristics participated in the same group or were groups to be made up of members with different characteristics. The third aspect was the size of each group. Several authors recommend a minimum of four to ten participants. This ensures that the discussions and the time for participants to contribute are not too limited (Russell, 2002; Ritchie, 2003).

With the assistance of the key informants, four focus groups each consisting of nine to sixteen people were formed. The groups were: seasoned banana farmers; general banana traders; women farmers who are also banana traders; CBOs, and NGOs working on HIV/AIDS issues in the area. Composition and selection of the group members was based on the topic that each group was to discuss.

The group of seasoned banana farmers was composed of tissue-cultured and non-tissue-cultured banana growing farmers (both men and women) who had grown bananas for a long time in the area. Although the discussions in each group were scheduled for one day, this group could not cover all the topics in one session and the discussions were done in two successive sessions. In the first session the group discussed the following issues: the local structural definition of household and the types of households in the area; major livelihood activities in the area; history of banana production in the area; their perceptions on the role of banana as a food security crop in the area; comparison of tissue-cultured and conventional banana in terms of initial costs, perceived risk, and saving in time and effort, increase in production and overall farm income. They also discussed the compatibility of tissue-cultured banana with existing cultivation practices. During the second session the discussions focused on the gender aspects in banana production in terms of labour profile in banana, division of labour and changes that occurred with the introduction of tissue-cultured banana. The group also discussed issues on decision making with regard to which part of the land unit to grow which crop, how much to grow and which banana varieties to grow. They also discussed gender differences in access to productive resources such as land, manure, fertiliser, water etc., access and control of cash generated from banana production and any changes that have occurred with the introduction of tissue-cultured banana. Their views on HIV/AIDS impacts on agricultural production and especially tissue-cultured banana adoption and production in terms of labour, cash generated, and access to land especially for widows, care of orphans, and abandonment or poor management of orchards were also elicited.

The banana traders' group was divided into two groups. The first group consisted of women banana traders, while the second one was made up of both men and women traders. The second group was formed after the realisation that although marketing of banana is predominantly a woman's activity in Maragua, it emerged during discussions with key informants that men had recently got involved in the trade (see Chapter 4). The discussion in the two groups centred on: market outlets for bananas; constraints in marketing banana; differences in marketing of the tissue-cultured and the conventional banana; use of income earned from sale of banana; control of income earned from sale of banana; their perceptions on the relationship between their trading activities and susceptibility and vulnerability to HIV/AIDS.

The fourth group consisted of the representatives of the National AIDS Control Council, HIV/AIDS related community-based and non-governmental organisations (CBOs and NGOs respectively) working in the area. This group discussed the trends of HIV/AIDS prevalence rates, effects on the community's livelihood activities and outcomes, coping strategies adopted by affected households and the community and possible areas of intervention.

A fifth group composed of extension service providers in the area, was scheduled but the turn-up for the meeting was very poor, making it impossible to hold a discussion. The information for this group was collected through in-depth interviews with the government extension agents and key informants. I was also involved in meetings organised by Ministry of Agriculture (MOA) extension agents that brought together stakeholders in extension service provision. Table 3.1 shows the composition and number of participants in the focus group discussions.

I moderated all the focus group discussions with the help of a research assistant who also took notes and audio-recorded the discussions. Two local extension officers (male and female) were involved in the discussion as facilitators. The discussions, which took from two to three hours, were held in the local coffee cooperative boardroom, which was a familiar venue to most participants.

Table 3.1. Participants in focus group discussions.

Focus group	Male	Female	Number of participants
Seasoned banana farmers	7	4	11
General banana traders	2	9	11
Women farmers/ banana traders	2	11	13
CBOs and NGOs working on HIV/ AIDS in the area	5	11	16
Total	16	35	51

3.3.4 Household survey

A survey questionnaire was administered to a non-probability sample of the population. The questionnaires were administered face-to-face as this provided the opportunity for further probing. The head of the household, as designated as such by the members of the household, was interviewed. Data was collected on household demography, assets, resources, labour and activities, crop production, banana farming and the suitability of tissue-cultured banana and effects of HIV/ AIDS on household assets and livelihood activities.

Sampling procedure

Initially the sampling procedure was set up as a stratified random sample, involving four categories of farming households: HIV/AIDS affected non-tissue-cultured and tissue-cultured banana farming households; non-affected HIV/AIDS non-tissue cultured and tissue-cultured banana farming households. However, for several reasons it was not possible to follow this procedure. The main reason was that it was difficult to get a list from which a sampling frame could be drawn. When I visited the District Planning Office in search of the sampling list used for the frequent surveys that the office is involved in, the person in-charge informed me that he was not authorized to give out the list to people not attached to their ministry. Efforts to get a list of farmers from the Divisional Agricultural Extension officer (DAEO) were fruitless, as the office did not have a complete list for the whole division or even the study area. The only available list from the DAEO's office was a list of farmers in two focal areas where they had introduced tissue-cultured bananas in the previous year as part of their extension programme.

The other reason was that it was difficult to identify households that were HIV/AIDS affected due to the stigma attached to the disease. A list of names of people, who died of AIDS, provided by the chiefs of the two locations, was of little help because it was difficult to locate their former residences. Finally two key informants, a person living with HIV/AIDS and a member of a community group involved in a voluntary feeding programme for AIDS orphans and their families, introduced me to affected families, which included both tissue-cultured and non-tissue-cultured banana growing households. Because of these difficulties a multi-stage sampling approach was used that involved several sampling methods, ensuring that each of the originally identified categories was sufficiently included. Purposive sampling was used to select some of the HIV/AIDS-affected households. Using names of affected households provided by the chief and the key informants. Participants were chosen on the basis of the presence of not less than 50 stools of conventional or tissues culture bananas which is equivalent to 0.08 hectares. The key informants then introduced me to these households and explained why they had been chosen for the interview. This made it relatively easy to elicit information on HIV/AIDS status of the household.

Tissue-cultured banana farming households were also purposively sampled. Using the list provided by the DAEO's office, participants were selected from some of the farmers that had been involved in the KARI tissue-cultured banana dissemination project and those that had adopted the technology thereafter through the extension programme.

To increase the numbers in all four categories of the sample convenience sampling was used. This involved visiting a certain area of the location, dropping the interviewers at different points from where they were to select households in one direction, skipping three to five households in between. Some households sampled in this category were classified as HIV/AIDS-affected based on proxy indicators developed on the basis of preliminary data analysis. According to these the HIV/AIDS status of households was based on the following variables: if any member of the household was experiencing illness due to tuberculosis, pneumonia, meningitis and/or typhoid; if the illness had lasted for more than six months or had been recurrent; if a household member had passed away during a period of five years prior to the time of survey and the death was caused by any of the above diseases. Another variable was having members that had either moved in or out after the death of an infected relative. The last variable was households with children who had stopped going to school. These included both secondary and primary school pupils because although the government has recently offered free primary school education in Kenya there is still the cost of school uniforms and books which some households cannot afford.

Based on this sampling procedure a total of 254 households were interviewed, distributed over the four categories as shown in Table 3.2.

Table 3.2. Sample composition of the household survey.

Using tissue-cultured banana technology	HIV/AIDS status		
	Affected	Non-affected	Total
Yes	21	46	67
No	54	133	187
Total	75	179	254

Training of interviewers

Initially it was planned that the two local agricultural extension officers who had been involved in identification of key informants and in the focus group discussions would be used as interviewers in the household survey. However, this was not possible as they had to go for in-service training. With the help of key informants I sourced interviewers from among the local people. Thirty local young men and women were interviewed and twelve of them selected. The criteria for selection were:

level of education where the minimum qualification was a high school certificate; working experience with a local NGO in the community; fluency in spoken and written English; good knowledge of the local language; and good reputation in the local community.

The training, which took three days, involved explaining goals and objectives of the study, imparting specific skills to be used in interviewing and motivating the interviewers. The training also included reaching a consensus among the interviewers as to the right interpretation of questions since the interview involved translating the questionnaire into the local language. During the training the need for the interviewers to reassure the participants of confidentiality in disclosing sensitive information was emphasized.

The interviewers worked in pairs of male and female where one would be asking the questions while the other would record the answers. The gender mix was in anticipation of households where respondents would not be free to be interviewed by either a male of female interviewer, but such cases were not encountered. It was also meant as a security measure for the interviewers as the area was reported to be insecure. The team selected a team leader who was responsible for collecting all the questionnaires at the end of the day and reporting any difficulties encountered during the day. I went through questionnaires noting mistakes, which were then discussed with the team the following day before they started working. The questionnaire was pre-tested for two days with a group of forty-eight households that formed part of the study. Pre-testing was necessary to determine the strengths and weaknesses of the survey concerning question clarity, format, wording, flow, order and timing. After pre-testing the researcher went through the questionnaires with the interviewers to find out whether the questions were comprehensible to both the respondents and the interviewers. All the unclear questions were either reformulated in a language that was understandable for both the interviewers and respondents without changing the meaning of the original questions. Two interviewers were discontinued in the early period of the exercise when I realized that they had too many unanswered questions in their questionnaires.

Some respondents refused to be interviewed citing the many surveys that had been conducted in the area without having seen their benefits. Others wanted to know whether it was a government development programme and if participation in the survey would ensure their involvement in the programme. Thus the interviewers had to clearly explain the purpose of the research. It was explained to the respondents that the survey was being carried out as part of an academic study and not as a baseline study in preparation of a bigger project. However, they were promised that the findings and the recommendation of the study would be made available to policy makers who might use them in future development projects in the area.

3.3.5 In-depth interviews

In-depth interviews were used to validate and clarify the information collected in the survey and in the FGDs. The selection of households for in-depth interviews was based on the results of the preliminary analysis of the quantitative survey. Six households were selected from the four categories mentioned above. The in-depth interviews entailed asking questions, listening to and recording the answers, and then posing additional questions to clarify or expand on a particular issue. Russell (2002) lists four types of interviews: informal interviewing, unstructured interviewing, semi-structured interviewing, and structured interviewing. In this study, semi-structured interviews were used. A list of questions and topics based on the outcomes of the quantitative survey and the focus group discussions were used to serve as a checklist during the interview. Although this checklist guided the interviews, respondents were encouraged to express their perceptions in their own words. The interviews also provided a forum for respondents to speak openly about private or sensitive issues that could not be discussed in a group setting or during the survey. They also served to reveal divergent experiences and 'outlier' attitudes, as groups often do not allow you to see experiences that may vary from person to person. During the interviews, which lasted from 30 to 60 minutes, more details were sought on gender division of labour, access to and control of assets, and details on impacts of HIV/AIDS on affected households.

We audio-recorded the in-depth interviews and this allowed me to devote my full attention to listening and probing the interviewee. The audio recording was also advantageous as it provided an accurate verbatim record of the interview.

Figure 3.2 shows the mix of methods used in data collection and the contribution of each method to the study.

3.3.6 Secondary data

A profile of the study area was done through collection of secondary data on various aspects, including geographic and demographic characteristics, institutional infrastructure, social structure, agricultural activities, awareness of HIV/AIDS, prevalence rates and national and local mitigation measures being taken. This information was collected from government central statistics office, country household survey reports, local government statistics such as district agricultural reports and participatory rural appraisal reports. Other sources included World Bank reports, UNAIDS reports, grey cover reports by national and international non-governmental organisations, local universities, in-country research and consultant teams.

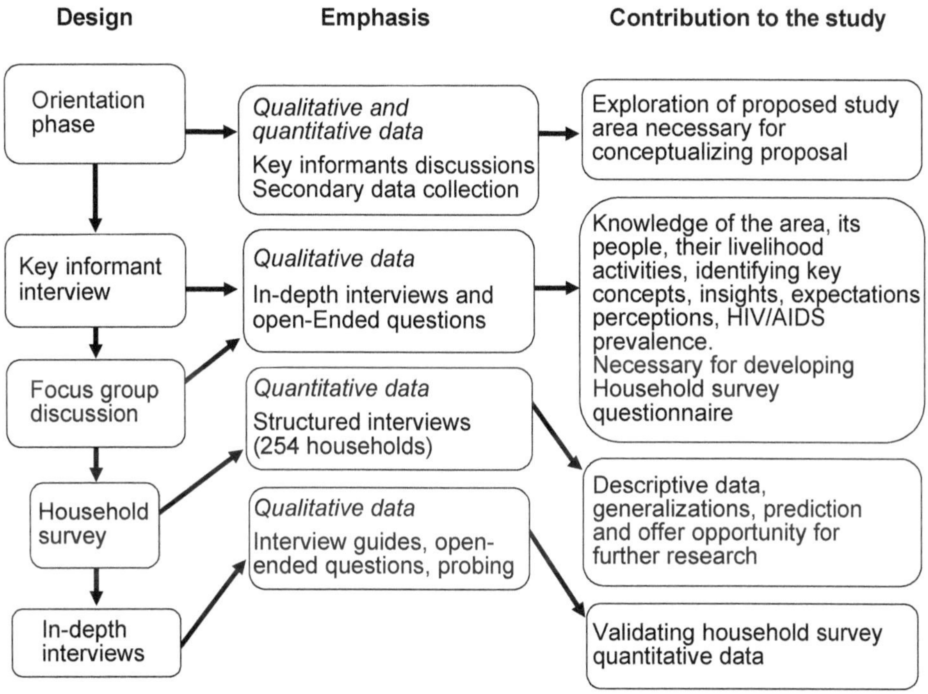

Figure 3.2. Method mix used in data collection.

3.3.7 Ethical considerations in data collection

This research raised ethical issues that are worthy of attention. It has been argued that for research in HIV/AIDS issues to be ethical, it must go beyond description and analysis to social responsibility for action (Ringheim, 1995). One of the questions that almost every participant in the study asked was: 'How are we going to benefit from all these questions that you are asking us?' This was despite the fact that the objectives of the study had been clearly explained at the beginning of the interview. This prompted me at the end of the study to collect food and clothing from family and friends to distribute among HIV/AIDS orphans and their guardians during one of the feeding days organized by one of the community-based organizations.

Another ethical issue that was encountered was that of confidentiality. HIV/AIDS is still highly stigmatized in spite of the high level of awareness that has been created through the media, NGOs and CBOs. The need for confidentiality was a high priority for most of the affected households which was emphasized during the interviewers training. The interviewers were provided with identification cards, which had their name, national identification number and the KARI logo. At the back of the card the interviewer had also signed to keep the information provided confidential. This helped the respondents in building confidence in the interviewers. Interviewers were also made to understand that household members were only to be interviewed on a voluntary basis, and indeed there were cases where the team was not welcomed. Households that were positively identified by the community and health workers as affected were either interviewed by the researcher or the research assistant.

A third ethical issue involved monetary incentives. The participants in the FGDs were compensated for their travel costs and lost wages incurred in attending the meetings. The amount compensated was based on the prevailing daily wages for farm labourers.

3.3.8 Problems encountered in data collection

As already mentioned above, the fear of stigmatization leads to silence in affected households. This presented a big challenge in identifying HIV/AIDS households during the study. The use of proxies to indicate affected versus non-affected households may also have led in classifying households as affected while they were not and *vice versa*. This is despite having been directed to affected households by social workers and key informants because in most cases the afflicted members did not admit their status. However, households that were taking care of AIDS orphans were more willing to talk of the departed members than households who had a member suffering from HIV/AIDS.

Another problem involved questions that required the respondents to recall events in the past. These included questions such as the amount of produce harvested, quantity sold or consumed, and amount of money spent on hospital bills medication and funerals.

A third challenge was the interpretation of the questions from English into the local language. Though the interviewers were thoroughly trained before starting the survey, clarity could not be guaranteed.

3.4 Data processing and analysis

3.4.1 Qualitative data analysis

Key informant interview notes were analysed by hand while audio records from focus group discussions were transcribed verbatim. Preliminary analysis of these data was used to fine tune definition of concepts based on the emic perspective which was necessary in formulating the household survey questionnaire. The focus group discussions and in-depth interviews audio records were further manually analysed and interpreted through the production of descriptive and explanatory accounts.

3.4.2 Analysis of survey data

The quantitative data was processed through manual editing and coding and eventually was analysed using SSPS. Before analysis, index variables (composite variables) were constructed to enable comparison of independent (tissue culture, non-tissue culture, HIV/AIDS affected/non-affected) and dependent (assets, livelihood activities, and outcomes) variables. For HIV/AIDS proxy indicators were formulated to distinguish affected from non-affected households. The total household asset value was computed by adding up the value of each tangible asset owned by the household. These assets included household items (bicycle, handcarts, radio, television, etc.), farm implements, farm structures, and livestock.

Using the SPSS statistical package the data was subjected to bivariate and multivariate analysis which provided descriptive statistics and associations between and among variables.

3.5 Time schedule of the research

I began this study in May 2003. The first year (May 2003 to April 2004) I spent in proposal writing and following relevant courses at Wageningen University. In year two (May 2004 to July 2005) I conducted the fieldwork in Kenya. I worked on data analysis, synthesis and wrote up the thesis in the third and fourth years.

Chapter 4

Kenya and the study area: agriculture and HIV/AIDS

In this chapter, the role of agriculture in the economy, the new agricultural policy changes being introduced by the government and the role of agricultural technological innovations are discussed. A brief discussion of the agricultural extension services is made. Subsequently, banana production and the introduction of the tissue-cultured banana technology are presented. Banana marketing and gender issues in marketing are discussed. The HIV/AIDS situation in Kenya and the interrelations between agriculture, gender and HIV/AIDS are explored. Finally a general description of the livelihood profile of the study area and the participants' perceptions of the HIV/AIDS situation in the area are presented.

4.1 Agriculture in Kenya

4.1.1 The significance of agriculture for the economy

Agriculture has been described as the cornerstone of human life and the backbone of many economies in the developing world particularly in sub-Sahara Africa. Strong agricultural growth has been a feature of countries that have successfully reduced poverty such as India, Bangladesh, Indonesia and China. Thirtle *et al.* (2003) reporting on observations drawn from 48 developing countries, show that a one percent increase in agricultural productivity reduced the proportion of people living on less than US$1 per day by between 0.6 and 1.2 percent. According to Lipton (2001) no other sector offers the same possibilities to create employment and lift people out of poverty.

Agriculture is the lifeline of the 61 percent of Kenya's population who live in rural areas. It is the root of economic growth, employment, and foreign exchange. Kenya's agricultural sector directly influences overall economic performance through its contribution to GDP. The sector contributes to 30 percent of the GDP, 80 percent of the national employment, 60 percent of the total export earnings and provides for most of the country's food requirements. Furthermore, agricultural growth has the potential to catalyse growth in other sectors, with an estimated multiplier effect of 1.6, compared to 1.2 in non-agricultural sectors (Republic of Kenya, 1999).

However, for the sector to effectively contribute to national growth and reduce poverty it needs to grow at about 4-6 percent per annum. On the contrary, this has not been the case as over the years the sector has experienced a continuous fall in growth from 4.7 in 1996 to 1.8 percent in 2004 (Republic of Kenya, 2003a). Several factors have contributed to this dismal performance: (1) unfavourable micro-and macro-economic environment, (2) inadequate markets and marketing infrastructure, (3) unfavourable external environment, (4) inappropriate legal and regulatory framework. Other constraints include (5) inadequate financial services, (6) inadequate storage and processing capacity for perishable commodities, (7) weak and ineffective research-extension-farmer linkages, (8) poor coordination with other support sectors such as water, roads, energy and security. There are constraints which are almost entirely exogenous, including (9) unfavourable weather conditions and high dependence on rain-fed production and (10) population pressure on the natural resource base (Republic of Kenya, 2005a). Lastly, (11) the impacts of HIV/AIDS on human capital and other assets is threatening to reduce any gains that may have been achieved over the years.

Notwithstanding this poor performance and the constraints enumerated above, the government continues to rely on the agricultural sector as the leading driver for development in Kenya. The Poverty Reduction Strategy (Republic of Kenya, 2001a), and the recently launched Economic Recovery Strategy for Wealth and Employment Creation (Republic of Kenya, 2003a) both identified agriculture as the sector that will steer economic recovery, create wealth and employment for the Kenyan people. The sector is expected to play this role through the strategies spelt out in the Strategy for Revitalizing Agriculture (SRA) launched in 2004, which aims at 'transforming Kenya's agriculture into a profitable, commercially oriented and internationally and regionally competitive economic activity that provides high quality gainful employment to Kenyans' (Republic of Kenya, 2004b: 2). The strategy proposes to provide a policy and institutional environment that is favourable to increasing agricultural productivity, promoting investment, encouraging private sector participation in agricultural activities and agribusiness. Agricultural research and the resulting technological innovations are one of the key components recognized in the strategy that are expected to contribute towards achieving the primary objective.

4.1.2 Agricultural technological innovations

It is proposed that technological change that leads to increases in food production and or reduction in production costs can reduce poverty in four basic ways: (1) raising the incomes of farmers who adopt the technology, (2) changing demand for agricultural labour, (3) reducing food prices (or dampening food price increases), thus making incomes purchase more and, (4) possibly stimulating economic growth that may generate additional employment opportunities and increase wages (Kerr & Kolavalli, 1999). The impacts of agricultural technology are demonstrated by the Green Revolution, which led to a doubling of yields for the major food grains in the 1960s and 1970s, particularly in Asia. The greatest impact was increase in crop production, which contributed to rural employment and lowered food prices (De Janvry & Sadoulet, 2002; Tripp, 2001a). In South Asia performance of high yielding varieties (HYV) was especially impressive, where between the mid-1960s and the mid-1980s, wheat yields increased by 240 percent and those of rice by 160 percent (Kerr & Kolavalli, 1999). Modern varieties have been adopted far beyond the bounds of the traditional Green Revolution area (Byerlee, 1996). Despite the scanty agricultural statistics from sub-Saharan Africa a number of success stories have been noted. These include the widespread adoption of new varieties of maize, wheat and rice varieties. Besides the modern varieties of grains, new varieties have also been developed for non-cereals principally for resistance to pests and diseases. These include improved cassava varieties in West Africa, new disease-resistant bean varieties for Eastern Africa, and new varieties of potato, among others (Maredia *et al.*, 2000). Crop management technology has also made significant contributions to increased agricultural productivity. This has been in areas such as soil fertility management, reduced tillage and use of integrated pest management (IPM) methods.

Biotechnology has great potential to develop crop technologies with favourable attributes such as higher yields, higher nutrient content, and resistance to pests and diseases. The adoption of genetically engineered crops has resulted in agricultural productivity growth and ensured an abundance of food in countries like the USA, where they have been adopted (Fernandez-Cornejo & McBride, 2002). However, biotechnology needs to be directed towards addressing the crops and areas that will benefit the poor people. Currently biotechnology is being developed by the private sector and is directed towards areas and crops that have promising returns (Tripp, 2001b).

Despite decades of investment in new agricultural technology, hunger and poverty continue to plague large areas of the developing world. This is particularly true in Sub-Saharan Africa where the impact of technologies has been less apparent. For agricultural technology to effectively contribute in raising productivity and poverty reduction, a set of interventions are also needed. These include secure output markets, effective supply systems (including credit), secure and equitable access to land and supporting infrastructure such as roads, telecommunication and irrigation (Dorward *et al.*, 2004). At the household level, the focus of this study, availability of resources and assets, intra-household relationships, social relationships, institutional support such as access to extension services, credit and markets, are all crucial for agricultural technology to make an impact to livelihoods of farming households.

Enhancing the adoption of appropriate agricultural technologies and practices is one of the imperative themes identified in the SRA for improving agricultural sector productivity and competitiveness in Kenya. The government proposes to restructure the agricultural research systems to enable it to address responsive and efficient technology development and transfer by: (1) increasing budgetary allocation for agriculture research to 2 percent of the GDP by 2010, (2) accelerating commercialisation of research products including contracts, and royalties for sustainability, and (3) involving stakeholders in research priority setting. The strategy also proposes to restructure the extension service to respond to user demands through: (1) partially privatising the extension service to compliment public services, (2) encouraging stakeholder participation in service provision, (3) facilitating and promoting capacity building of extension service providers, and (4) developing performance standards and monitoring evaluation frameworks for extension services (Republic of Kenya, 2004b).

4.1.3 Agricultural extension services

There is a general consensus that extension services, provided they are well designed and implemented promote agricultural productivity providing farmers with information that helps them to optimize their use of limited resources (Evenson & Mwabu, 1998; Bindlish & Evenson, 1993). Nevertheless, agricultural extension services are faced with challenges of providing relevant agricultural extension and training programmes to meet farmer-needs in a changing agricultural environment. Globally extension services are facing challenges emanating from the changing social and natural environments as well as from within the extension organisations themselves (Leeuwis & Van den Ban, 2004). These authors discuss challenges related to ensuring food security for a growing world population, reducing poverty and promoting agricultural practices that ensure sustainable natural resource management. Other challenges include upcoming new agricultural technologies and new clientele - the elderly and orphans managing farm household due to HIV/AIDS effects on the household structure (Du Guerny, 2002).

The Government of Kenya has a long history in providing agricultural extension services, dating back to the colonial period. The extension service in Kenya has operated under two major systems. The first is the government extension system. Under this system several extension models and approaches have been tried, including the model farmer approach, the integrated agricultural rural development approach, farm management, training and visit (T&V), the farming systems approach and the farmer field schools. These methods have produced varying levels of achievements (Muyanga & Jayne, 2006). The second type of extension system is the commodity-based systems run by government parastatals, out-grower companies, and cooperatives. The commodity-based extension deals mainly, but not exclusively with commercial crops such as coffee, tea, pyrethrum and sisal, the system is motivated by profits. The system integrates all aspects of producing and marketing a particular commercial crop ranging from research, advice, input supply to farmers, to organizing local and exports marketing.

The extension services in Kenya, like in many other developing countries, are particularly constrained by insufficient human and financial resources occasioned by the SAPs in the 1980s. Budgetary allocations for extension services have declined from 6 percent of the overall annual government budget in the two decades after independence to less than 2 percent currently. About 20,000 government extension workers would ideally be needed to respond to farmers' needs countrywide, but only 7,000 are currently employed (Republic of Kenya, 2005a). Apart from financial constraints agricultural extension is currently facing challenges related to providing services to HIV/AIDS-affected households. Traditional extension programmes aimed at male household heads are not likely to reach orphan- and female-headed households or even the elderly. Leeuwis and Van den Ban (2004:11) recommend that for agricultural extension services to support farmers in dealing with these challenges they will have to be 'reinvented as a professional practice; that is, it will have to significantly adapt its mission, rationale, mode of operation, management and organisational structure'.

Approaches to agricultural extension service provision

Currently extension services in Kenya are guided by the National Agricultural Extension Policy (NAEP) formulated in 2001 (Republic of Kenya, 2005a). This policy aims at making the national extension system more responsive to the needs and realities of its clients. This is to be achieved by promoting decentralization of decision-making, promoting linkages between the ministry and other stakeholders in agricultural extension and increased community involvement in extension programmes. The policy is implemented through the National Agricultural and Livestock Extension Programme (NALEP), supported by the Government of Kenya and the Swedish International Development Agency (SIDA). This new approach is intended to promote transparency and accountability in the agricultural extension service.

The NALEP programme is currently focusing at the location rather than the division level which was the focus of previous extension programmes. before. In the new approach the location (with between 2000 – 6000 small-scale farmers) is divided into four working blocks and targeted for a period of not less than one year for extension services. The community draws a Community Action Plan (CAP) which is supposed to pave the way for the entry of various stakeholders. Common Interest Groups (CIGs) are formed around enterprises with good opportunities in the area. The CIGs are expected to be self-driven and to seek extension services that they need from government or private providers. The role of the government extension officer is to provide training in crop and livestock management aspects, linking the CIGs with sources of inputs, marketing information and other service providers. They also supervise and coordinate the activities of various service providers to protect the interests of farmers. Resource-poor and vulnerable farmers who do not join any of the CIGs are targeted individually (Republic of Kenya, 2004a). In spite of these policies that are generally supportive of a shift toward demand-driven services, with a diminished role for government, the challenge that still remains is to put them into practice.

4.2 Banana production in Kenya

Banana and plantain contributing to food security in developing countries are an important source of nutrition, and are estimated to supply more than 75 percent of the carbohydrates needed by more than 20 million people. Banana is fourth on the list of the developing world's most important food crops after rice, wheat and maize. Sub-Saharan Africa produces 35 percent of the world's bananas and plantains. East Africa (notably the Great Lakes region covering portions of Uganda, Rwanda, Burundi, Tanzania, Kenya and Congo) is estimated to produce 15 million tons per year with the highest per capita consumption in the world (INIBAP, 2000).

Kenya produces 510,000 metric tonnes of bananas on 40,000 hectares of land (FAO, 2006). Bananas are grown from sea level to 1800 meters above sea level, although production is concentrated in wetter areas. Cultivation is largely carried out under rainfed conditions and is primarily done by small-scale farmers. The average banana holding is 0.3 hectares (Qaim, 1999), although nearly every rural household has at least a couple of banana plants around the homestead. Both cooking and dessert varieties are produced in different parts of the country and their distribution is influenced by factors such as local taste, eating habits and marketing considerations. The East African highland cooking bananas (AAA)[2] and the Apple banana (AB) are grown in Nyanza and Western Provinces. In Central, Eastern and Coast Provinces the Cavendish (AAA) predominate although local varieties such as *Muraru,* in Central and Eastern Provinces *Bokoboko,* in Coast province, also exist (Nguthi, 1999).

Data from FAO (2006) shows that although production in terms of area planted and the quantity produced has increased, the actual yield per hectare has slightly

[2] Plantains and bananas *(Musa* spp. L.) originated from intra- and interspecific hybridization between two wild diploid species; *M. acuminata* Colla. and *M. balbisiana* Colla. The two species contributed the A and B genomes, respectively. Various *Musa* genomic groups consist of either one or different permutations of the basic genomes. Thus, cultivated dessert and East African Highland bananas are classified mainly as AAA, plantains are AAB, and cooking bananas are ABB. Other combinations including AB, AAAB, AABB, and ABBB also exist, which occur naturally or are produced by artificial hybridization (Stover and Simmonds, 1987).

declined over the years (Figure 4.1). This is mainly due to increasing incidences of pests and diseases.

Banana farming in Maragua

During focus group discussions and in-depth interviews it emerged that farmers have grown bananas in Maragua for many years as a staple food alongside coffee as their main cash crop. Traditionally farmers produced local varieties such as *Muraru, Kiganda, Mucuru* and *Gacukari* (Apple banana) mainly for their own consumption. However, from the late 1970s the price of coffee started going down and coffee co-operative societies could no longer provide farmers with inputs like fertilizers and chemicals for control of pests and diseases, leading to decline in production and quality. By the mid-eighties coffee production became increasingly unsustainable for small-scale farmers and many uprooted or intercropped coffee with other crops such as beans and maize. It was then that banana production gained prominence as a sustainable enterprise with unique advantages. Unlike coffee which is solely an export crop, bananas could readily provide food for the family as well as cash. The crop provided a regular source of income throughout the year as opposed to coffee where payments were made once a year. By the mid-1990s what started as a substitute crop to coffee mainly grown by women for food, banana became the main cash crop in Maragua.

However, this situation was threatened with the onset of fusarium wilt, a fungal disease caused by *Fusarium oxysporum f. cubense,* a soil borne fungus that proliferates

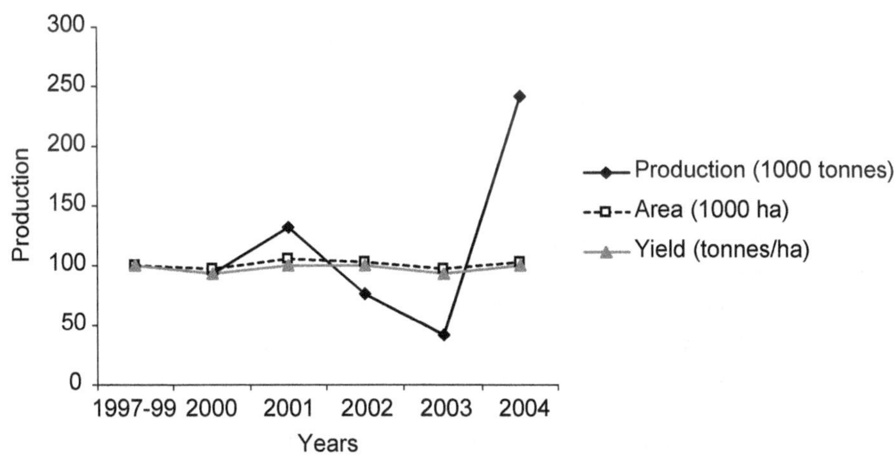

Figure 4.1. Banana production trends in Kenya over the last five years. Source: FAOSTAT (2006).

in the soil and can persist in the soil for over 30 years. The disease was first reported in Australia in 1876 and by the 1960s it had spread to most of the major banana-growing regions of the world (INIBAP, 2000). In Kenya it was first observed in 1952 but did not cause any damage to bananas until in more recent years (Kung'u, 1998). Today the disease is prevalent in most banana growing regions and especially so where susceptible cultivars such as Gros Michel, Apple banana, *Muraru* and *Bokoboko* are grown. Until the mid-1990's, Gros Michel was the predominant commercial cultivar grown in Maragua. By 1997 fusarium wilt disease had devastated bananas in Maragua and in some instances caused a hundred percent yield losses, leaving the farmers deprived of their source of livelihood (Kung'u, 1998). Control of the disease has been attained in other banana growing countries by the replacement of Gros Michel with other tolerant varieties, notably the Cavendish and the FHIA (Honduran Foundation for Agricultural Research). These varieties had been introduced to farmers in Maragua by KARI (Kenya Agricultural Research Institute), but their adoption was limited by among other things the unavailability of clean planting material. The introduction of the tissue culture technique for rapid propagation of clean planting material in 1997 was thus perceived as having the potential to help reverse the declining trends of the crop.

The Kenya Agricultural Research Institute (KARI) in collaboration with the Rockefeller Foundation (RF) and International Development Research Centre (IDRC), engaged in a project for the production and delivery of clean banana planting material to smallholder farmers in the country. The International Services for Acquisition of Agri-biotech Applications (ISAAA) facilitated the project with technical backstopping from the Institute for Tropical and Sub-tropical and Crops (ITSC) of South Africa. The overall goal of the project was to make available to small-scale resource-poor farmers clean and improved banana seedlings that would raise yield at farm level and contribute to food security and poverty reduction.

4.2.2 The tissue culture technology

Plant tissue culture refers to growing and multiplication of cells, tissues and organs on solid or liquid media under aseptic and controlled conditions. The commercial technology is primarily based on micropropagation, in which rapid proliferation is achieved from tiny stem cuttings, auxillary buds, and to a limited extent from somatic embryos, depending on the plant species in question, in suspension cultures and bioreactors. The prefix 'micro' generally refers to the small size of the tissue taken for propagation, but could equally refer to the size of the plants that are produced as a result. In practice, the 'explant' is removed surgically, surface sterilized and placed on a nutrient medium to initiate the mother culture, that is multiplied repeatedly by subculture. Tissue culture techniques are used throughout the world as a tool both in research and in commercial production. Some of the important applications include: mass production of uniform plants often referred to as clonal propagation, elimination of pathogens from plant tissues to obtain clean plants, induction of genetic variability and improvement of crops through genetic engineering. The technique allows mass multiplication of species that are difficult to regenerate by conventional methods of propagation such as the banana. Cultivated bananas are parthenocarpic fruits (i.e. do not produce seeds) and subsequent crops are usually produced from suckers of the mother plant. Conventionally, one banana plant only produces five to ten suckers in a year depending on the variety. Traditionally, planting materials for bananas (in form of suckers) is usually exchanged freely or at a small cost among farmers. This is a simple cost-free practice but has a major disadvantage in that many of the banana pests and diseases are distributed through this exchange. According to FAO (2001) tissue culture has revolutionized banana cultivation and has replaced the use of conventional vegetative suckers in many of the intensive banana-growing regions. FAO estimates that up to 50 million tissue-cultured plants are produced annually making banana the most widely *in vitro* propagated plant. One shortcoming of the technique is that it does not eliminate all of the banana viruses.

In developed countries the tissue culture techniques have been used to a great extent for banana, but in developing countries the techniques have not as yet been fully exploited. This is mainly because commercial development with tissue culture has focussed on the Cavendish clones on which the international export trade depends and very little has been done for other varieties. In Kenya tissue-cultured banana technology was introduced in 1997 to curtail the declining yields of banana due to pests and diseases. The technology was introduced as a package, which included the tissue-cultured plantlets, the Cavendish disease tolerant varieties, and information on crop husbandry and post harvest handling practices (Wambugu & Kiome, 2001). An *ex-ante* study by Qaim (1999) showed that the tissue-cultured banana technology was likely to raise yield by 150, 132, and 93 percent for small, medium and large-scale farms, resulting in increases in incomes by 156, 145 and 106 percent for small medium and large scale farmers respectively. Currently three laboratories (one private-GTL and two institutional- KARI and JKUAT) are producing tissue-cultured banana.

4.2.3 Banana marketing

In Maragua, banana marketing has always been primarily a woman's activity. During the early 1980s the town council had allotted a central market where the stop-over farmers (mostly women) would bring their bananas for sale. The market was a popular stop-over for buses travelling between Nairobi, Central and Eastern Kenya for commuters who would buy ripe bananas. However, when the council introduced a levy on produce being brought into the market, most of the farmers stopped bringing their bananas to the market. But since a demand had already been created and there were women who were involved in banana marketing along the roadside as a livelihood activity the trade continued. The women traders started going to buy bananas from the farms and farmers preferred this because they could sell their produce and still continue doing their work on the farm. About 86 percent of the interviewed households sell their bananas when green at the farm level to middlemen/women bulk buyers. At the farm level, there are no fixed prices for bananas and the selling price is usually negotiated between the seller and buyer depending on the size of the banana bunch. The farmer has to bargain to get the best price for his produce, ranging between Ksh.100 – 250 ($US 1- 4). This is one of the reasons why women than men are more involved in selling. Participants of the focus group discussions said that most men have no bargaining skills and will usually sell at whatever price the buyer offers, especially if the buyers are experienced market women. Women on the other hand have bargaining skills and know how to deal with their fellow women seriously. According to one of the participants, 'the notion that women are understanding or merciful is only true with their children but not in trading or business issues'. Women are also more informed on the prevailing prices because they are the ones who go to the market. The farmers prefer to sell their bananas at the farm gate due to the high transportation cost of delivering a few bunches to the market.

After buying from the farm the traders harvest the bananas thus the farmer does not incur direct harvest cost and associated losses. The trader transports the bananas to Maragua trading centre using hired handcarts, bicycles or pick-up trucks, depending on the volume of the produce. The women traders have formed groups of three or four who hire a room in Maragua trading centre that they use both for ripening, storing and as distribution centres for their bananas. The bananas are ripened in big metal drums into which passion fruit; avocado or banana leaves are put. These materials produce ethylene gas, naturally emitted by fruit, which is essential for the ripening process. The bananas take two three days to ripen. When the bananas are ripe they are either sold locally in the trading centre, to buyers who come from other rural trading centres in the area or from Nairobi. The bananas are sold in hands of five or six fingers depending on the variety and the quantity bought. According to the traders the new Cavendish varieties (locally known as Nyoro) have bigger bunches but smaller fingers than Gros Michel (locally known as *Kampala*). Consumers prefer Kampala because of its longer fingers and attractive colour after ripening. However, the Cavendish varieties are sweeter but take longer to ripen and are more susceptible to bruising. The fact that the two varieties ripen at different times is good for business as the period of selling is extended. However, Gros michel is being replaced by the Cavendish varieties because of the former's susceptibility to *Fusarium* wilt.

A marketing initiative was introduced with the intention of organising banana growers in Central Kenya into a marketing association. The association was launched in February 2003 under the name Highrigde Banana Growers' and Marketing Association (HBGMA). The association was provided with seed money by ISAAA to help in starting up its activities. The membership consisted of group members of eight self-help groups that had been growing tissue-cultured bananas under the ISAAA/KARI tissue-cultured banana project in the region. The membership was to be enlarged to include other banana growers in the country. The HBGMA's main objective was to promote production and marketing of bananas through the establishment of produce collection and market information centers. The association also intended to establish distribution centers for tissue-cultured plantlets and other production inputs such as agrochemicals (fertilizers and herbicides).

The farmers expected that by coming together they would strengthen their bargaining capacity and avoid exploitation by middlemen, thus increasing their income. They also expected that a good market would lead to increase in production and income. This rise in income would then enable them to invest into other income generating activities. However, these expectations have not been realized.

In Maragua there were two groups: Nginda and Ichagaki. Soon after the launch of the association, the group leaders acquired office premises and an office worker and both were to be paid using the seed money provided by ISAAA. Members in both groups started to contribute Shs.100 per month. They started marketing their bananas through the Kenya Horticultural Exporters (KHE) but soon this market outlet was lost due to poor quality and low quantities. This, plus leadership wrangles among the association's committee caused members to lose faith in the association and they stopped contributing money in their groups and many discontinued membership altogether. Finally the two groups merged to form one group, which currently has about 30 active members.

Although other market outlets have been offered to the association, members have been reluctant to sell their bananas to them. This is because they are offering lower prices than they get in the local markets and they end up with large volumes of rejected produce due to low quality. The rejected bananas cannot be easily sold in the local market, which buys whole bunches as opposed to hands which the formal marketing agents prefer. Thus the farmers have stopped marketing bananas as a group and have gone back to individual marketing.

Virtually all the informants interviewed indicated that the association was still relevant and its objectives valid. Some of key informants felt that the association needed to be nurtured for a longer period through the provision of financial and logistical assistance. During the nurturing period members should be educated and trained to a point where they gain ownership of the association. They feel that strong, visionary leadership accompanied by accountability should be instituted.

Gender issues in banana marketing

Literature points to the fact that women in Africa are increasingly becoming food providers as well as cash crop farmers (Spring, 2000). In Maragua it is apparent that women have access to banana money but variations were observed in the control of income from bananas in different households as indicated in the case studies below.

The case of a banana husband farmer and a banana trader wife

Florence's husband is in-charge of the farming activities while she is involved in marketing. They own a big farm (>5 ha) most of which is planted with banana. Her husband delivers the green bananas from the farm to her ripening room in Maragua. She ripens and sells locally as well as to buyers from other towns. She also hires young men who sell bananas at a commission along the Nairobi-Murang'a highway. At the end of the day she buys the household items that are required such as food and clothing and gives the rest of the money to the husband. There is usually transparency and agreement on what to do with the proceeds. The couple has educated their children, built a permanent house and even bought a pick-up truck which the husband uses to transport the bananas.

The case of a de facto female household head banana farmer

Nyambura is a *de facto* household head as the husband is working and living away from home. She is therefore in charge of all the farming household activities, including the sale of produce from her four hectare farm. She is growing tissue-cultured banana and sells the produce at farm. She makes decisions on how to spend the money but she must communicate these to the husband. She spends the money to hire farm labour, buy inputs, food stuffs, and household items such as clothes for herself and the children. The husband's salary is used to pay for school fees and other major expenses.

The case of a pensioner banana farmer and a banana trader wife

Muriuki, a pensioner, works in the banana farm while his wife is a banana trader. According to Muriuki, when a banana buyer comes, it is the wife who goes with him/her to the banana plot because she is the one who goes to the market and therefore knows the prevailing prices. After she has sold the bananas in the *shamba* (plot) she comes and lays the money on the table and since there is no other source of income for the family they then sit down and decide what to do with the money. But she must first of all bring the money to the table. The only way she can cheat the husband is if she fails to say the exact amount for which she sold the bananas.

She also sells bananas in Maragua town where she has a ripening room. They are not paying school fees as all their children have finished school and left home. She sometimes buys bananas from her husband as well as from other farms. She uses this money for household items and food stuff.

The case of husband and wife who have divided the banana farm
Janet and Kamau have divided the banana farm into two. Janet harvests from a smaller portion of the banana farm and she sells to get money for household spending such as sugar, salt, food items, clothes, etc. She also buys from other farms to augment the sale quantities and she is in control of this income. Kamau, on the other hand, sells the bulk of his produce at the farm level, especially to meet major household expenditures such as education of the children. He sells the bananas while green at the farm and will usually stick to one buyer with whom he has an arrangement whereby the buyer pays in advance, so that he is able to meet large payments such as school fees. This is because as he cannot get enough bananas to sell at one time, as they mature at different times, to raise the money needed to pay for school fees. The buyer then tags bunches he has paid for and harvests them at his convenience.

The case of a banana farmer husband and a banana trader wife
Njoroge's household is composed of his wife and three young children aged between ten and three. He works on the farm where he is growing tissue-cultured banana, maize, bean and Irish potato on their 0.4 hectare piece of land. His wife is a banana trader in Maragua town. She recently started selling bananas in Nairobi where they have rented a stall in one of the markets. But due to her household chore demands this activity does not look viable and they are thinking of quitting. Njoroge's wife is in control of the money from banana sales. She buys all the household items and food stuff required. When Njoroge requires money, which is rare since he has very few needs as he does not take alcohol, he gets it from his wife.

These case studies indicate that women in Maragua seem to have a large measure of control over banana income. The women said that banana production is different from coffee where although the woman provided labour for management and harvesting, she neither knew how much was paid nor had any access to or control of the money. This is because the money was paid to the male household head either through the bank or the coffee society.

4.3 HIV/AIDS situation in Kenya

Since the first case of HIV was diagnosed in Kenya in 1984, Kenya has gone through different phases of national response. The first phase in the late 1980s and early 1990s was characterized by widespread denial of the existence of HIV/AIDS in the country.

During these early years of the epidemic political commitment was limited until 15 years later when former President Mr. Moi declared HIV/AIDS a national disaster, which was his first major public statement on the subject. By then, an estimated one in every nine sexually active persons in the country was already infected. This was followed by the adoption of a multi-sectoral policy in the mid-1990s which saw the establishment of a national policy in the Sessional Paper No. 4 of 1997 on AIDS in Kenya (Republic of Kenya, 1997). Since then the government has shown strong political will and support in the fight against HIV/AIDS, demonstrated by the current president's personal involvement and his declaration of the 'Total War on HIV/AIDS' in March 2003. Presently there is a massive increase in the number of actors and resources to support the national response.

4.3.1 The HIV/AIDS pandemic in Kenya

Levels and trends of HIV/AIDS

The 2003 Kenya Demographic and Health Survey gives Kenya's estimated HIV/AIDS prevalence in adults (15–49 years) as 7 percent (with a range from 6.1% to 7.5%) and just over 6 percent (5.2-7.0%) in 2005 (UNAIDS, 2006). This is much lower than the 15 percent estimate by UNAIDS in 2001 (UNAIDS, 2002). The number of people living with HIV/AIDS in Kenya includes about 1.1 million adults between 15 and 49 years, 60,000 aged over 50, and approximately 100,000 children. The prevalence rate of women is almost twice that of men. HIV infection among adults in urban areas (10%) is twice as high as in rural areas (5%). There are differences in regional prevalence, with Nyanza province having the highest (15% in adults), while North Eastern Province has the lowest (less than 1%). Information from sentinel surveillance indicates that adult prevalence peaked at a level of 10 percent in the late 1990s. However, the annual number of AIDS deaths is rising steeply, and has doubled over the past six years to about 150,000 deaths per year. This is attributed to the large number of people who were infected in the 1990s. Providentially, the number of new infections each year has dropped from 200,000 in the 1990s to around 80,000 at present (Republic of Kenya, 2005b). Table 4.1 shows HIV prevalence in the country by gender, residence and province.

Institutional policy response

The first government response to the epidemic was the establishment of an AIDS control Committee in 1987 in the Ministry of Health. This committee developed the first and second 5-year strategic plan for AIDS control (1987-91 and 1992-96). The plan emphasised the need for creation of awareness about AIDS, blood safety, clinical management of AIDS and capacity building for the management of AIDS control programmes at the national level. The second medium term plan (1992-

Table 4.1. HIV/AIDS prevalence by gender, residence and province in Kenya (2004).

	Women	Men	Total
Total	8.7	4.6	6.7
Age			
15-19	3.0	0.4	1.6
20-24	9.0	2.4	6.0
25-29	12.9	7.3	10.4
30-34	11.7	6.6	9.4
35-39	11.8	8.4	10.1
40-44	9.5	8.8	9.1
45-49	3.9	5.2	4.4
50-54	n.a	5.7	n.a
Residence			
Urban	12.3	7.5	10.0
Rural	7.5	3.6	5.6
Province			
Nairobi	11.9	7.8	9.9
Central	7.6	2.0	4.9
Coast	6.6	4.8	5.6
Eastern	6.1	1.5	4.0
North Eastern	<1.0	<1.0	<1.0
Nyanza	18.3	11.6	15.1
Rift Valley	6.9	3.6	5.3
Western	5.8	3.8	4.9

Source: Republic of Kenya, 2005b.

1996) stressed the need to involve all sectors in HIV prevention in order to mobilise broader national response against the epidemic. Soon after this the government launched the Sessional Paper No. 4 of 1997 on AIDS in Kenya (Republic of Kenya, 1997). This is the policy document that provides guidance on national response. The National AIDS Control Council (NACC) located in the Office of the President, was established in 1999 to provide leadership and coordination of the different sectors involved in the national response to HIV/AIDS. Donor funds to fight AIDS (mainly from the Global Fund) are channelled through NACC. The Council came up with a HIV/AIDS Strategic Plan for five years (2001-2005) with the main objectives of reducing prevalence among youth by creating AIDS control units which would mainstream AIDS into ministry activities and creating constituency AIDS-control committees (CACCs) to represent a wide range of stakeholders at all levels.

These objectives are to be realized through five priority activities: prevention and advocacy; care and treatment; impact-mitigation; monitoring and evaluation; and management and coordination. The Council is broad and includes all the government ministries from the provincial to constituency level, civil society and development partners. It is run by a secretariat which serves as the Council's executive arm. The strategic plan is having a positive impact. Indeed, Kenya is the second country in sub-Saharan Africa, after Uganda, to register a sustained decline in national prevalence levels. Although part of the decline in prevalence rate is attributed to deaths from AIDS-related diseases, there is also evidence that it stems from changes in Kenyans' sexual behaviour, expanded HIV information campaigns and community involvement, voluntary counselling, testing programmes and improved access to anti-retroviral therapies. For example, The Demographic Health Survey (Republic of Kenya, 2003b) reported that 24 percent women are using condoms during sex with casual partners compared with 15 percent five years earlier. In addition, the proportions of men and women with more than one sexual partner reduced by more than half in 1993-2003, and more young men and women are delaying their first sexual encounters (Republic of Kenya, 2005b).

The government has recently formulated a new Kenya National Strategic Plan (KNASP) 2005–2010, which will be used to scale-up the national AIDS response using the 'Three Ones' principles. The 'Three Ones' are guiding principles for national authorities and development partners in the fight against AIDS that were identified through a preparatory process at global and country levels initiated by UNAIDS (UNAIDS, 2004). They are: an agreed HIV/AIDS action framework that provides the basis for coordinating the work of all partners; a national AIDS coordinating authority, with a broad based multi-sector mandate; and an agreed country-level monitoring and evaluation system. The new KNASP sets three key priority areas namely: prevention of new infections; improvement of the quality of life of people infected and affected by HIV/AIDS; and mitigation of socio-economic impacts of HIV/AIDS. This study hopes to contribute to the latter.

Despite the positive steps taken in policy formulation and the reduction in HIV/AIDS prevalence several challenges still impede the fight. These include new infections among the youth, high death rates, high cost of ARVs, discrimination and stigma, and lack of harmonisation of activities and resource mobilisation. In relation to the high cost of ARVs the government recently waived charges levied on anti-retroviral drugs in government hospitals and health centres (Daily Nation Friday, June 02, 2006).

4.3.2 Agriculture, gender and HIV/AIDS in Kenya

A simulation study of the potential impact of HIV/AIDS for different sectors in Kenya carried out by USAID (Bollinger *et al.* 1999) projected that agriculture would suffer the greatest impact because the largest share of the labour force in the country works in agriculture. According to this study AIDS will lower the value of agricultural production between 1.7 percent and 2.4 percent by 2010. Although the prevalence rate has gone down since this study was conducted, the data still indicates the potential impact of the epidemic on the agricultural sector.

Although the micro-level empirical evidence on the effects of HIV/AIDS on agriculture is still quite limited, the weight of the evidence to date does indicate that AIDS is having a measurably adverse impact on agricultural production in Kenya. A study conducted to investigate the impacts of HIV/AIDS on commercial agriculture in Kenya (Rugalema, 1998) showed significant social and economic impacts in the agro-estates. There was a significant increase in medical and funeral expenses incurred by the agro-estates, loss of productivity due to absenteeism as a result of workers' morbidity and funeral attendance, and loss of highly trained and skilled and experienced employees. In another study conducted to assess the efficiency of tea estate workers who had died or were medically retired because of HIV-related causes between 1997 and 2002, Fox *et al.* (2004) described the labour productivity decline associated with HIV/AIDS-related morbidity. They found that a tea plucker who eventually died of an AIDS-related condition was absent from work 31 days more often, spent 22 days more on light duty per year. The person also produced an average of 7.1 kg less tea leaf per plucking day than a person dying from non-AIDS-related causes.

At the household level Yamano & Jayne (2004) found that the death of an adult female caused a greater decline in the cereal area cultivated, whereas cash crops and non-farm income were most adversely affected in households incurring a prime-age male adult death. The relatively high prevalence rate among women in Kenya worsens the situation because women supply 70-75 per cent of agricultural labour (Spring, 2000). These consequences are compounded by the fact that women have limited access to productive resources such as land and credit. If a husband dies, the wife's lack of collateral limits her ability to obtain credit to continue with the farming activities. Also, an increase in the number of orphans increases the child-care responsibilities of healthy women in the community who must care for the sick as well. These additional duties multiply the negative impact of AIDS on agriculture because these women will have less time for farming activities.

4.4 Livelihood profile of the study area

4.4.1 Human capital

Maragua district has a population of 409,229 people (based on census projections for 2002) and it is one of the most densely populated districts in the country with 488 persons per square kilometre. The average household size is seven persons.

Provision of public education and health services originate from macro-level policies designed to raise the level of human capital across the country as a whole. The government recently introduced free primary education in Kenya which will substantially contribute to meeting the MDG goal of universal access to primary schooling by the year 2015. However, the infrastructure expansion and human resources implications of the free primary education policy are immense, even with the financial support from the World Bank and other donors (Republic of Kenya, 2003c). The enrolment, retention, completion and progression rates for boys and girls at primary and secondary levels are almost equal in Kenya. In Maragua district there are 251 primary schools, 80 secondary 13 youth polytechnics one teacher training college and a farmer training centre. The gross enrolment rate in primary and secondary education is 121 and 42 respectively with a 1:1 ratio of girls and boys (SID, 2004).

The district is served with one government hospital, five health centres and fourteen dispensaries. Maragua district hospital is a rural health training facility as well as an out and in-patient public hospital with a Comprehensive Care Centre (CCC) for the HIV/AIDS programme. The CCC began as a Voluntary Counselling and Testing (VCT) unit in 2002. It is a partnership initiative between the Ministry of Health (Maragua), WEM - Integrated Health Services (Thika), Medicos Del Mundo (Spain) and Association VIHDA. WEM Integrated Health Services is a local NGO based at Thika which has developed an effective model for mobilizing Community Orphan Care Committees. They address children's needs for nutrition assistance and material necessities like clothing, school uniforms, and shoes, as well as create support groups for people living with AIDS. Association VIHDA is an international NGO that assist people living with HIV/AIDS (PLWHA) and handles the medical ARV treatment at the CCC. Medicos Del Mundo is a Spanish NGO that runs the Prevention of Mother to Child Transmission program. They also run the VCT, train the medical staff on screening, diagnosis and treatment of STDs, counselling on STIs and HIV/AIDS, and have a condom distribution network.

The hospital, which has a clientele base from as far as the districts of Nyeri and Thika, provides several services at the CCC: provision of free anti-retroviral drugs to HIV patients, group/support therapy, voluntary counselling and testing, positive orphans therapy (some are on ARV treatment), and prevention of mother-to-child transmission. Some of the ART beneficiaries offer volunteer services in the hospital on a regular basis. At the time of the study there were about 670 enrolled patients at the CCC with approximately 260 on anti-retroviral therapy. According to data collected from the CCC, HIV/AIDS prevalence in the area is estimated at 6.1%. (NACC, 2003).

4.4.2 Natural capital

Maragua district covers an approximate area of 1065 km^2 and most of the land (77.7%) is arable. The average farm size is 0.93 hectares with a total of 48,747 farm families. The land in Maragua was demarcated in 1965 and farmers issued with title deeds. However, population pressure has led to sub-division of the land into smaller sizes which have not been registered. The study area lies in the main coffee zone and is characterized by small-scale mixed farming with pronounced banana farming. However, in the recent past coffee production and income has been declining due to low international market prices, mismanagement of co-operative societies, high cost of inputs and low productivity. Other crops like bananas, French beans and other horticultural crops are rapidly replacing it. Table 4.2 shows the area, production and value of some of the crops grown in the district.

Table 4.2. Crop production statistics for the year 2003 in Maragua district.

Crop	Area (Ha)	Total production (Tons)	Value (KSh M)
Banana	3,355	38,040	285
Maize	17,300	23,335	234
French bean	950	7,480	187
Bean	10,100	5,454	109
Coffee	8,726	1,800	70
Sweet potato	290	2,900	58
Irish potato	2,120	84,800	51

Source: 2001, District Agricultural office, Maragua.

4.4.3 Physical capital

The study area is served with a network of roads covering 1038 km but most of them are earth roads as only 13 percent are tar-marked. Most of these roads become nearly impassable during the rainy season, making it difficult for farmers to get their produce to the markets. Electricity is found in the trading centres and homes around and within coffee estate farms formerly owned by white settlers. These homes were supplied with electricity based on a policy that existed during the colonial period which required that all households within a radius of 0.5 km around coffee estates be supplied with electricity, probably to ensure security. Very few households have landline telephone connections but about 60 percent of the household have mobile connections with the two countrywide mobile service providers, Safaricom and Celtel. There is one post-office in Maragua town that serves the whole district. It offers banking services, postage of surface mail, delivery of goods and parcels, speed post services, telegrams, money orders, sale of stamps and daily address boxes. The post office also offers internet services and all the government district offices have access to e-mail services. The internet services are mainly utilized by the young generation i.e. high school and school leavers youngsters. The service is reliable and is only interrupted when there is no electricity.

Water

Over the years the Kenya government has taken various initiatives to improve access to safe clean water and had set a target to secure water for all by 2000. However, despite the government's efforts to achieve this goal, today only about 1 million Kenyans out of a total population of 33 million have access to clean water (Wambua, 2004). As a result, over half of the households in the study area use water from the river for drinking, cooking, bathing and washing clothes. There are six major rivers: Maragua, Irati, Sabasaba, Tamara, Githanja and Itherui, numerous streams and plenty of ground water which has been tapped by digging boreholes and wells. One-third of the households have sunk boreholes in their farms, while 13 percent source their domestic water from wells. A small proportion (7.5%) has piped water in the houses, yard or have access to paid or free tap water. This piped water is provided by the Kenya Railways water supply system which serves the local railway station. Households with piped water pay flat rates for their water services irrespective of the amount of water they consume.

Women bear the primary responsibility for water collection in Maragua. In a few cases men collect water but this is usually from the borehole, a well or a water tank located within the compound. On average, the daily number of trips made and the average time spent on collecting water per household are 2.3 and 20 minutes respectively. In households where the wife was mentioned as the main person who collects water, 52 percent of these women make between 1-3 trips per day. The majority of households (50.4%) obtain water from the river, a walking distance of 100 -500 m, while for a few households the distance to this water source is up to 5 km to fetch water. The time and energy that women and children have to spend on fetching water goes at the expense of economically productive agricultural production and income generating activities.

Fuel wood

In Kenya it is estimated that fuel wood accounts for 70-75 percent of annual energy consumption mainly as fuel wood for cooking and heating in rural areas and as charcoal in urban areas. Rural farming households often have few alternatives to fuel wood to meet their basic fuel need, and in many rural areas of Kenya, about 90% or more of the energy for household use as well as schools or clinics depends on fuel wood (Mahiri & Horworth, 2001).

A very small proportion of the households (4%) in the study population are connected to electricity, while 99 percent of the households use wood as their first source of fuel for cooking and heating. The use of gas is mainly found in wealthy households. The main source of fuel for lighting is paraffin although some households also use firewood. Two-fifths of the households use solar energy for lighting.

Although the district has 226 km^2 of forest land (Gatare and Kahumbu forest reserve), the forest is under the control of the government and cannot be used by households in the study area. Before the Land Adjudication Act in Kenya, fuel wood was collected free on communal land. However, land registration and transfer to private ownership coupled with increasing population pressure and the accompanying pressure on land leads to difficulties of collecting fuel wood (Mahiri & Howorth, 2001).

In the study area households have tried to solve the problem of fuel wood by planting *Grevillea robusta* trees (locally known as *Mubariiti*) along farm boundaries. Grevillea was originally introduced in the area as a shade tree for coffee plantations as it does not compete strongly with adjacent crops because it is relatively deep-rooted (Dewees, 1995). It tolerates repeated heavy pruning of its roots and branches enabling farmers to regulate the degree of competition with other crops. The tree branches are periodically harvested for firewood and the trunks used for timber or charcoal. Trimming is mainly done by men as trees always belong to them in spite of the women having put in substantial labour and management responsibilities for the production of the trees. Thus, even though there is an abundance of trees in the area the respondents reported a problem of fuel wood. This is because firewood collection is a woman's duty in 64%, of the households, but women cannot harvest firewood from trees without their husband's consent, which is actually a cultural and not an ecological issue.

4.4.4 Financial capital

The most important financial institutions existing in the area are rural banking institutions. Among the rural banking institutions Murata SACCO is the most active. It is owned by coffee, tea and dairy farmers through buying of shares. To become a member one has to buy a minimum of 30 shares at the rate of KSh. 100 (about US$ 1) per share plus another KSh. 100 membership fee. The bank provides ordinary savings account, fixed deposit accounts, school fees account, junior account, and a holiday or Christmas account. It also has a micro-finance scheme that provides short term loans to business people. This loan facility is administered through formation of self-help groups whereby individual group members are eligible to small loans after the group has saved for three months at a rate of 1.35 percent per month. The group has to be registered and members co-guarantee one another, the only collateral being group membership. Other short term loans are provided for school fees, to pay labour for coffee picking, tea picking and hospital bills. The bank also provides insurance services to cooperative societies, individuals and business enterprises.

Many informal micro-finance institutions are also operating in the area offering financial services to community groups. Unfortunately very few of these are involved in agricultural activities.

4.4.5 Social capital

Social capital, indicated by membership in community organization in this study, appears to be high when we consider the number of groups in existence. According to reports from the MOA there are about 154 women groups in the district whose key objective is a rotating savings-and-credit association (ROSCA). There are many burial associations that are clan-based. Most (99%) of the population is Christian, the rest are either Muslim or traditionalists. One of the traditional religious sects common in the area is *Mungiki* which means 'united people'. The sect is secretive, so their beliefs or origins are unclear. However, they favour a return to African traditions and reject Western culture. They support female circumcision and they have been linked to ethnic violence and anti-government resistance.

Various faith-based community organizations are involved in community-based activities. For example, the Catholic Church runs an orphanage and feeding programme for orphans while Presbyterian Church of East Africa (PCEA), has a training polytechnic for marginalized girls of the society. There are other community-based organizations (CBOs) that have been formed with the sole purpose of HIV/AIDS awareness creation (Table 4.3).

In focus group discussions with HIV/AIDS-related community-based organisations (CBOs) it transpired that though a lot of money had been given out to various organizations in the district, the funds have been poorly managed or ended up in the wrong hands, while the genuine beneficiaries continue to suffer. The participants felt that the government needs to control and vet the community groups being formed to support the orphans and being funded by NACC as most of them are not genuine and are misappropriating the funds. They also felt that funding should now shift from awareness creation to direct funding to the infected and affected people since most of people have been sensitized on HIV/AIDS prevention.

Table 4.3. Community based organisations (CBOs) contacted in Maragua.

Group name	Membership no.	Activity
Maragua Doop	30	AIDS awareness and better care of orphans
Maragua Single Mothers Women Group	30	AIDS and better care
Emilio and Good Hope	45	HIV awareness
Mwitindia Women Group	60	AIDS awareness
Vision Centre for Orphans	40	AIDS awareness and training
Urumwe Self Help Group	30	Teaching programs for orphans and the destitute
HEGA Herbalist	31	HIV awareness
Manjoice Mwihotori	20	Herbal practice Research and nutritional awareness
Macas Group	30	HIV/AIDS awareness and orphans better living
Gachaba Unity	25	Prevention and advocacy
Maragua Orphans	20	Helping orphans
Huruma S.H. Group	30	Helping orphans and tree planting
Nyakagumo Widows and Ushuhunda Mothers	20	Tree planting

Source: Focus group discussions, 2004.

4.5 Community's perception of the HIV/AIDS situation in the study area

The general level of health among the households in the study area appeared to be low. Ninety-three households out of 254 reported having had ill members over the last one year. Sixty-five households reported having one member who was not in good health, 14 households reported two members, 12 reported three, and two had more than three ill members. The most commonly mentioned illnesses were: malaria, typhoid, diabetes, tuberculosis, pneumonia and meningitis. Only six households openly acknowledged HIV/AIDS as the cause of illness of household members. Most of HIV/AIDS-affected respondents (of whom the interviewers were informed of their status) in the in-depth interviews did not admit that HIV/AIDS accounted for the ill health of themselves or other household members. This indicates that HIV-related stigma and discrimination is widespread. Stigmatisation was particularly prevalent in households that had recently lost a member or that had a sick member at the time of survey. In contrast, people freely discussed HIV/AIDS (locally referred to as *mukingo*) in focus groups, where it emerged that HIV/AIDS is widespread in the area. Indeed one of the groups ranked HIV/AIDS as the first killer disease in the area, followed by malaria, diabetes, typhoid and water borne diseases. One woman in a focus group discussion said: 'We are all affected; there is no household that can say they are not experiencing the effects of the disease.'

The origin and spread of HIV/AIDS in the study area was explored through focus group discussions, key informant and in-depth interviews. Maragua town used to be a stop-over point for trucks travelling to parts of Eastern and North Eastern Kenya in the early eighties when the main road passed through the town. As a result several slum areas cropped up in the town, such as Mathare, Moscow and others, which were notorious for prostitution. A case vividly remembered by the old participants in one focus group discussion is of an influx of Somali drivers who used to transport maize from the Cereals and Produce Board Maragua depot in big trucks. It is claimed that all the drivers' girlfriends of the time have since died of AIDS.

According to focus group participants several social trends were associated with an increasing spread of HIV/AIDS in the rural areas. Rural-urban migration was also identified as a determining factor in the spread of HIV/AIDS in the area. Migration disrupts marital and familial ties, leading to new networks of sexual partners. Migrated men still have ties and sexual relationships in their rural homes and often support the remaining members of the household. They eventually return to the rural homes when they are affected and thus increase the possibility of spreading HIV/AIDS to the rural areas. According to one man who operates a bar in Maragua town: 'Migrant workers from Mombasa, Eldoret, Nairobi and other towns come over during Christmas festivities and patronize the bars and in the process move around with women from the area'.

Increased pressure on natural resources due to population increase and lack of employment is a major factor that has created hopelessness and despair in the rural areas. Sub-division of land and continued cultivation with little replenishment over the generations has resulted in uneconomical land units, which can hardly support production for the households' livelihood. A young strong person who finds he has no land or property to inherit from the family and has no gainful employment, becomes desperate and starts engaging in any form of ill acts. This hopelessness prevails not only in the youth but in the general populace. One activity popularly quoted during the focus group discussions was engagement in drinking illicit brews and using drugs. As well as derailing rational decision making among users, heavy drinking and drug use also lead to promiscuity. The illicit brew users also fail in their roles as husbands and may force their wives to engage in sex work to fend for their families.

Decline in moral values was another factor associated with the spread of HIV/AIDS in the area. Among the young generation (particularly girls) exposure to the western culture and the peer pressure to embrace and conform to this lifestyle leads to prostitution since there are no what they would consider appropriate jobs (white collar-jobs) in the rural areas. As Mary an elderly widow said:

The culture has changed and people have adopted the western culture, which has reduced morals. The present generation is lazy and does not want to work on the farm but still they want to live a good life. Even here in the rural areas young people have been exposed to the western culture and because they do not have money to adopt this lifestyle they go out and involve themselves in sexual activities.

Some of the young people also do not believe in the existence of HIV/AIDS. An elderly man noted:

> *When you tell them there is AIDS (mukingo) they do not take it seriously. Even when somebody dies and it is very well known that he died from AIDS they say 'Gutiri kia blackie', meaning that ones life and death has already been predetermined and no one dies without God's will. For such a person it means he will not take any precaution even if he knows that so and so has died of AIDS. Such people are many, especially among the youth. There is a great problem even if we are being encouraged to talk openly about HIV we shall still talk so that he who will hear will hear.*

Degeneration of moral values is not only observed among the young but also among older married couples. This is because when an AIDS-afflicted husband comes from Nairobi and dies at home the wife who is left behind continues to have sexual relations with men from the local area or village and in the process spread the disease. A study conducted in various parts of the country reported that traditional values and norms related to sexual behaviour have eroded (Ocholla-Ayayo, 1997).

There are also cases of individuals who have been tested HIV-positive, but are still carelessly engaging in sexual activities and saying '*Nii Ndiri mutino*', meaning that the person is not liable to accidents or misfortunes and therefore does not need to be precautious.

Marriage customs among the young generation are also changing due to their poor economic status that may not enable them to pay bride wealth as well as sustain a family. The young men argue they would rather buy a cow that will bring some income than pay bride wealth for a woman who is going to be troublesome. Such young men will not fail to look for sexual pleasure with other married or unmarried women even though they are not willing to enter into a marital relationship.

The use of condoms is still not popular among the rural folk, especially so among the married women. In one focus group discussion the women reported that if they ever suggested to their husbands to use condoms they would be categorized as prostitutes. Some women fear that if they insist upon condom use or even ask their partner to use a condom, they will endanger their relationship, perhaps even lose their partner. And in losing their partner they may lose a relationship which confers status, emotional and, perhaps, financial support. Married women with dependent children are especially vulnerable. This demonstrates the limited bargaining power that women have in matters relating to avoiding risk in sexual relations.

Box 4.1. Degeneration of morals

As Mary said 'Before it was not so bad because it was usually the young unmarried people who were dying and therefore not leaving partners in the village, but now when a married man dies and leaves a wife it is dangerous for the community. It is not uncommon for a wife to die, leaving a husband who other women want to get married to in order to 'share the things that were left' without bothering to know what killed her. Increased access to contraceptives has also encouraged promiscuity amongst women where they no longer fear conceiving outside marriage with resultant HIV/AIDS spread'.

Stigmatization

Stigma is defined as 'prejudice, discounting, discrediting, and discrimination directed at people perceived to have AIDS or HIV, and the individuals, groups, and communities with which they are associated' (Herek, 1999). AIDS stigma is manifested in discrimination, violence, and rejection of people with AIDS. The stigma is characterized by silence, fear, and denial. Stigma fuels the spread of HIV/AIDS. It undermines prevention, care and support thus increasing the impact of the epidemic on individuals and families. HIV/AIDS related stigma is associated with behaviours that may be illegal or forbidden by religious or traditional teachings, such as pre- and extra-marital sex, and sex work. It is also drawn from HIV/AIDS' association with some of the most fundamental parts of the human experience: sex, blood, disease and death. Stigma is also associated with conditions that are perceived to be contagious. People fear the risks posed by casual contact with people living with HIV/AIDS. Particular behaviours such as prostitution are also considered disgraceful, and persons practicing prostitution are similarly stigmatized. The early association of HIV/AIDS with prostitution, combined with the reality that most cases of AIDS in Kenya are acquired through sexual transmission, fuels stigmatization.

However, the existence of the Maragua VCT and Comprehensive Centre is helping in reducing the stigma by creating awareness. The Centre holds monthly meetings for PLWHA during which they are encouraged to share their experiences. Those who have been engaged in the group for long and have disclosed their HIV status have received counselling and training and have now become trainers. According to focus group participants active involvement of PLWA awareness creation is central to the fight against stigma and reducing the spread of the disease. Figure 4.2 presents participants perception of the origin, causes of spread and consequences of HIV/AIDS in the study area.

Box 4.2. HIV/AIDS stigmatization

Annah belongs to Gacaba Women Group that is involved in a feeding program for AIDS orphans. They contribute food for the orphans (mainly maize and beans), which they distribute to families that are supporting AIDS orphans. But this season they have decided to buy bedding for the orphans, as they realized that due to the stigma associated with these children their relatives were not providing bedding and instead the children were sleeping on gunny and polyethylene bags. In some households the children are not allowed to share cups, plates, spoons or washing basins with other family members for fear of infection.

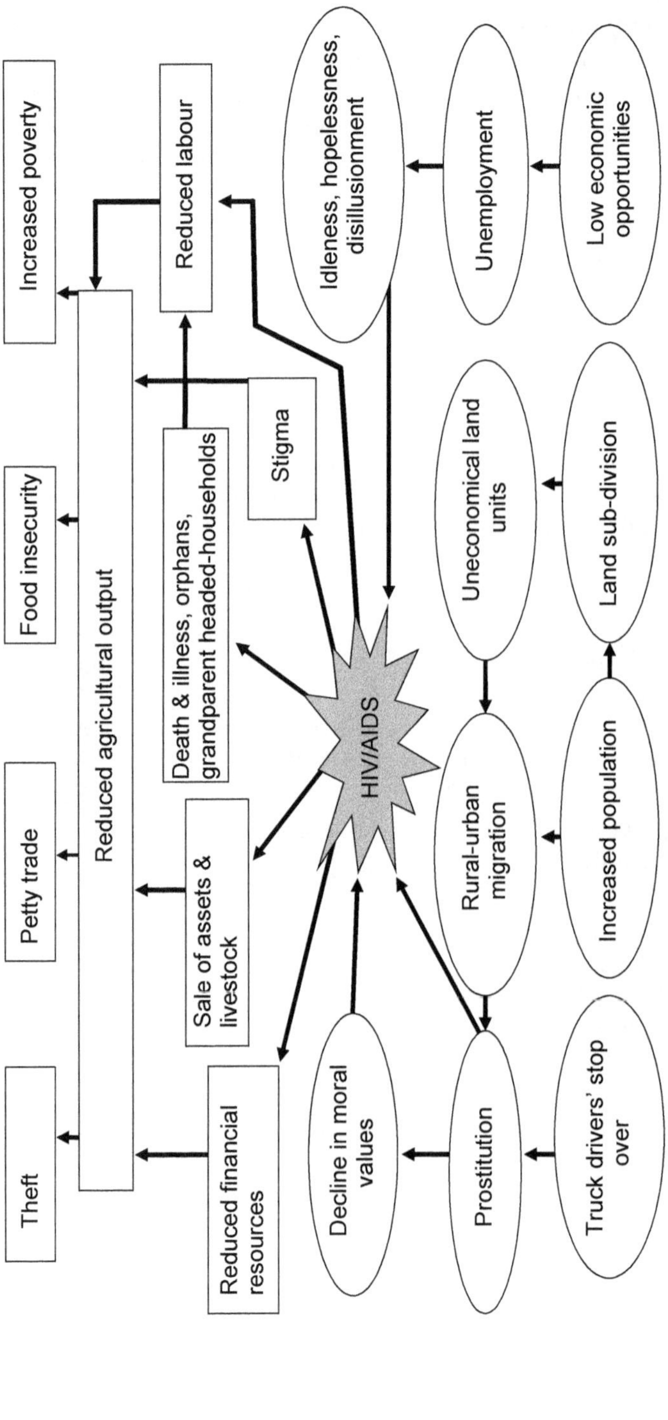

Figure 4.2. Participants perceptions of the origin, causes of spread and consequences of HIV/AIDS in Maragua.

Chapter 5

Household assets, activities and strategies

This chapter is presented in three sections. In the first section farm households in the sample population are characterised and classified into three comparatively homogenous categories using factor and cluster analysis procedures. In the second section a description and discussion of the resulting farm types is given. This discussion is organised around the five capitals of the DFID livelihood pentagon (DFID, 2001). Finally a discussion of the livelihood strategies and activities of the farming household types is presented. The discussion is done drawing on qualitative village-level data as well as quantitative household-level data.

5.1 Characterization of farm households

Due to the heterogeneity of the farm households in terms of assets endowments, the livelihood activities and the strategies they adopt differ. These differences also considerably influence households' responses to various interventions and shocks to their livelihoods. Thus there is need to characterize and classify farm households into sets of homogenous groups with similar characteristics to study their access to resources, response to constraints and to various development interventions in their livelihoods. There is a wide scope of literature on farm household characterization using various criteria. For example in an *ex-ante* study conducted to assess the impact of tissue-cultured banana in Kenya, Qaim (1999) characterized Kenyan banana producers into small, medium and large farms based on the size of land under banana. Tittonell *et al.* (2005) used production activities, household objectives and the main constraints faced by farmers to characterise farms in assessing nutrient depletion and soil degradation in Western Kenya. Salasya (2005), in a crop production and soil nutrient management economic analysis of households in Western and Central Kenya, used variables related to management decisions, structural farm household characteristics and distance to the nearest market to classify farms in her study. Selection of variables used in this study was based on the five capitals of the livelihood framework. Factor analysis and the cluster analysis methods were used. Factor analysis is a group of multivariate statistical methods whose primary objective is to reduce a large number of variables to a smaller number of factors. It summarizes the information contained in a number of original variables into a smaller set of composite dimensions or factors with a minimum loss of information. Unlike other multivariate analysis methods where one variable is considered the dependent variable and all others independent, in factor analysis all variables are interdependent and considered simultaneously. It can be used for several purposes but in this study we used the method to select a subset of variables or factors from a larger set of human, financial and physical capital variables using a component pattern matrix (Field, 2000). Table 5.1 shows the variables selected in creating human, financial and physical capital factors in the sample population. The size of suitable agricultural land owned and membership in a community group were selected without being subjected to factor analysis to indicate natural capital and social capital respectively.

The next step was to identify relatively homogeneous subgroups within the sample population that were based on the five livelihood capitals that could be grouped into farming household types (FH) using cluster analysis. Cluster Analysis (CA) is a classification method that is used to arrange a set of cases into clusters which both minimize within-group variation and maximize between-group variation. Variables are classified such that each variable is very similar to others in its cluster with respect to some predetermined selection criteria (Hair *et al.*, 1987). Using the factors created in the factor analysis in the previous section cluster analysis identified three classes of farming households counting 111, 94 and 46 households respectively. Table 5.2 shows means and standard deviations of some of the major livelihood characteristics of the three farm types. The means of each farm type was compared with the combined means of the other two variables using a t-test.

Table 5.1. Variables used in creating human, financial and physical factors.

Factor	Variable
Human capital	Size of Household
	Age of household head
	Education level of household head
	Sex of household head
	Use of hired labour
Natural capital	Size of suitable agricultural land
Financial capital	Access to credit
	Shares
	Financial savings
Physical capital	Personal household items value
	Total livestock value
	Total farm equipment value
	Total household asset value
Social capital	Membership in a community group

Source: Household survey, 2004.

Table 5.2. Means and standard deviations of livelihood indicators for three farming household types in Maragua.

Variable	FH type one (N=111)		FH type two (N=46)		FH type three (N=94)		Overall (N=251)	
	Mean	SD	Mean	SD	Mean	SD	Mean	SD
Human capital								
Household size	5.4	2.1	6.2	3.2	5.8	2.3	5.7	2.5
Age of household head (years)	53.1	17	54.3	12.1	52.3	14.3	52.9	15.1
Educational level of household head[1]	2.1***	0.6	2.8	0.6	2.8	0.8	2.5	0.1
Sex of household head (1=male)	0.8***	0.4	0.9	0.3	1	0.2	0.9	0.3
Hired labour (1=yes)	0.3***	0.4	0.7***	0.4	0.8	0.4	0.6	0.5
Natural capital								
Farm size (ha)	0.5***	0.3	1.6	1	1.2	1.9	1	1.3
Land:labour ratio	1	2.2	1.4	3.4	1.1	1	1.2	2.8
Financial capital								
Household saving (1=yes)	0.2	0.4	0.9***	0.3	0.2	0.4	0.5	0.5
Access to credit (1=yes)	0.9	0.3	0.9	0.4	0.8**	0.4	0.8	0.4
Physical capital								
Total livestock value (Ksh '000)	36.3***	28.5	96.3***	59.8	66.1***	53.6	58.5	50.8
Total farm equipment (Ksh '000)	9.2***	19	60.9	67.7	72	131.5	42.2	91.1
Personal household items value (Ksh '000)	2.3**	5.1	24.7	77.3	44.8**	140.5	22.3	93.8
Social capital								
Group Membership (1=yes)	0.8	0.4	0.9	0.4	0.9	0.3	0.9	0.4

*, **, *** identifies variables that are significantly different at the 10%, 5% and 1% level respectively when the mean of variable is compared to the mean of the other two combined.

[1] education levels: 1 = none, 2 = primary, 3 = secondary, 4 = college/university.

Source: Household survey, 2004.

5.2 Description of asset ownership in different farming household types

It is apparent from the above table that farming households significantly differ in their asset endowment. These differences are discussed in the section below under the five livelihood capitals.

5.2.1 Human capital

Significant differences were observed among the three farming household types in the gender and education level of the household head and use of hired labour. Households in FH type one (N = 111) have a higher proportion of female heads with lower education level than FH type two and three (Table 5.3). Households in farm type one also have fewer members but a higher dependency ratio compared to type two and three. Farming household type one has a higher proportion of household heads in the 65 + bracket compared to type two and three. Households in farm type one have the largest proportion of widows, separated, divorced and single heads. The majority of households in farm type one never hire labour. Households in FH type two (N = 94) are mainly headed by men and have the highest number of heads with college or university degrees. This group also has the highest proportion of employed household heads. The majority of FH type three (N = 46) households are headed by males with a high level of education. They have the largest household sizes and hardly ever use hired labour. The majority of the household heads in this category are married and over one-quarter are employed.

5.2.2 Natural capital

A significant difference was observed in land size among the three farming household types (Table 5.2). Generally households own small pieces of land with an average of 0.94 hectares. Households in FH type one have small pieces of land while there is no significant difference in type two and three households (Table 5.4). Land renting in although not common is practiced in a few households. Renting in is more common in FH type one and three than in FH type two while renting out is common in farm type two and three which have bigger farm sizes. Households are renting in land to grow food crops and vegetable.

Discussions in focus groups with elderly banana farmers indicated that land is owned by senior males in the households and inheritance of the father's land is the most common mode of land acquisition. This was confirmed in the household survey where majority of the households in the three farming household types indicated they had inherited their land (Table 5.4). Traditionally the father divides the land according to the number of wives, sons or unmarried daughters he has. A

Table 5.3. Frequencies for human capital variables by farm type in Maragua.

Variable	FH type one (N=111)		FH type two (N=94)		FH type three (N=46)	
	N	%	N	%	N	%
Gender of household head						
Male	86	77.5	91	96.8	41	89.1
Female	25	22.5	3	3.2	5	10.9
Age of household head						
15-65	85	76.6	79	84	42	91.3
>65	26	23.4	15	16	4	8.7
Household size						
01-Mar	18	16.2	18	19.1	10	21.7
04-Jun	62	55.9	43	45.7	18	39.1
07-Sep	22	19.8	23	24.5	10	21.7
<9	9	8.1	10	10.6	8	17.4
Dependency ratio						
Low (0-50)	55	49.5	56	60.2	28	60.9
Medium (51-100)	33	27	17	18.3	12	26.1
High (>100)	26	23.4	20	21.5	6	13
Education level						
None	11	9.9	2	2.2	-	-
Primary	80	72.1	37	40.2	15	33.3
Secondary	18	16.2	35	38	25	55.6
College/university	2	1.8	18	19.5	5	11.1
Use of hired labour						
Yes	30	27	75	79.8	34	73.9
No	81	73	19	20.2	12	26.1
Marital status						
Married	87	78.4	88	93.6	42	91.3
Single	1	9	-	-	1	2.2
Divorced	1	9	-	-	-	-
Separated	4	3.6	-	-	-	-
Widowed	18	16.2	6	6.4	3	6.5

Source: Household survey, 2004.

few households indicated that they have been temporarily allocated land by their parents who are still the custodian of the title deeds. This practice is common among

Table 5.4. Frequencies for various land variables by farming household type in Maragua.

Variable	FH type one (N=111)		FH type two (N94)		FH type three (N=46)	
	N	%	N	%	N	%
Land size (ha)						
<0.4	71	64	26	27.7	1	2.2
0.4-0.8	24	21.6	25	26.6	12	26.1
>0.8	16	14.4	43	45.7	33	71.7
Mode of acquisition						
Inherited	92	82.9	65	69.1	30	65.2
Permission to use	6	5.4	1	1.1	1	2.2
Purchased	11	9.9	19	20.2	12	26.1
Inherited and purchased	2	1.8	9	9.6	3	6.5
Land tenure						
Own with title deed	53	47.7	56	59.6	34	73.9
Own without title deed	57	51.4	35	37.2	11	23.9
Own with title deed and part without title deed	1	0.9	3	3.2	1	2.2
Women have equal access to land						
Yes	77	69.4	72	76.6	34	73.9
No	34	30.6	22	23.4	12	26.1
Women's mode of access to land						
Marriage	34	30.6	22	23.4	13	28.3
Inheritance	37	33.3	49	52.1	11	23.9
Marriage and inheritance	40	36	23	24.5	22	47.8

Source: Household survey, 2004.

the Kikuyu where a married son is allocated land until when the father feels he is ready to permanently divide his land among his sons. In the mean time, the son can make improvements to the land although there is no guarantee he will retain the same plot.

There are households that have bought land and the majority of these are in FH two and three. Focus group discussions revealed that some of these are people that had moved from the higher agro-ecological zones of Maragua district due to population pressure and land scarcity to the lower and less populated zones of the district. The warm climate in these lower zones enables the production of food crops, like maize, beans and bananas, twice a year to as opposed to the cool higher areas with one season and lower yields. There are also households who sell their small pieces of land, and since the value of land is high in Maragua (KSh 150,000-300,000/acre) they are able to buy bigger pieces in the less densely populated areas of Eastern Kenya (Meru, Embu) or in the Rift Valley, where land is cheaper than in Central province and eventually some household members may migrate to these areas.

Land tenure

Most of the land in the study area is freehold private land owned by individuals. While land registration is supposed to eliminate the traditional concepts of land ownership and increase security of land ownership, the customary concept of ownership still prevails and land relations are still based on the customary law in the study area. In FH type one the majority of the households (51%) had no title deed to their land and most indicated that they had 'been shown' their piece of land by their parents with the latter still being the custodians of the title deed. This indicates that claims to land are still being made on the basis of customary law even where such land is registered. In FH type two and three, which include slightly more households that have purchased land, more than half of the households have title deeds to their land (Table 5.4). Farming household type three has the highest proportion of households with title deeds to their land. In-depth interviews revealed that the land registration process is time-consuming and costly and most households are not willing to invest in the process as long as the clan is aware of their land ownership status. Moreover, all land transactions on freehold land must be discussed with community elders to determine the different rights of the members of that household before they are approved by the Land Control Board. Writing on the effects of land titling in Kenya, Whitehead & Tsikata (2003:73) report 'that a gap has developed between the control of rights as reflected in the land register and control of rights as recognized between most local communities'. This is because despite the existence of registered titles, access to majority of land plots is either through inheritance, non-registered sales, or gifts. In their study on the impacts of HIV/AIDS on land rights in Kenya, Aliber & Walker (2006) also found that most of the land in the rural areas is registered in the name of someone other than the person identified as the household head, usually a parent of the household head. This has resulted to what they call 'inter-generation conflict over land'. According to Aliber & Walker (2006:714) 'the older generation (especially fathers) seems to be protecting their own longer term tenure and preventing unacceptable sales'. The younger generation on the other hand feels constrained by lack of formal land ownership as they cannot confidently invest in the land. Lack of title deeds also makes one vulnerable if the actual owner decides the user should leave. Third parties may also seek to take control of the land by force or intimidation.

Land ownership and gender

For a woman to inherit her husband's land among the Kikuyu there are certain customary conditions that have to be met, one being that all the customary requirements of bride wealth must have been settled. Until bride wealth is paid the marriage has no legal status, and women have no rights and can be made to leave the homestead at any time. As the men say, the women can be 'chased away if they do not work hard'. The second condition is that she must bear male children. One of the participants of the focus group discussion said: 'You cannot leave land to a woman who has no children as she might run away', meaning she might decide to get married to another man and in the process abandon the clan land or sell it. If a woman has no sons one of her co-wives' sons or a male from the clan will inherit the land, because it is always assumed that daughters will get married and leave the clan. A woman without sons can also 'marry' another woman who will bear children for her, hopefully sons, who will then inherit the land. Although focus group participants said this was an outdated practice, in-depth interviews indicated that it still occurs among the older generation. Divergent views were expressed by the participants concerning unmarried or divorced women's access to land. Some people said that in the past unmarried and divorced women would inherit land based on how well they had served in the home, in which case the woman would be receiving a repayment as a servant, by virtue of her service to the clan or family. In the event that her father or members of the clan thought she had not rendered good services in the home she would not inherit land. This practice is still being maintained in some households but some participants said that today unmarried women inherit land by virtue of being children of that home and are treated like sons. Other participants said that today fathers are 'officially' allocating land to their daughters. As one participant said: 'These days we are giving them (daughters) land because if we do not the boys will chase them away when we die'. However, some elderly male participants categorically said that only sons can inherit their fathers' land. Aliber & Walker (2006:715) recorded similar findings in Thika, Embu and Bondo districts of Kenya where they observed 'an intergeneration shift in attitudes toward greater tolerance of the possibility of daughters inheriting land from their parents'

In most cases widows continue to farm land that was registered in their husband's name after his death, but with recent increases in land demand and value this security is no longer guaranteed as it is not uncommon for a son to sell the land without his mother's permission. Three categories of women's access to land use were identified as: access through marriage, inheritance, and both marriage and inheritance. The categories were identified during focus group discussions, where access through marriage meant that only married women in that household had user rights. Daughters are still expected to get land from where they are or would be married. Access through inheritance meant that married women had access to land use and could inherit the land after the death of their husband. It also meant that unmarried or divorced daughters as well can inherit land from their fathers. However, this last sentence is contentious as pointed out above. The third category was composed of households who indicated that women can gain access to land through both inheritance and marriage.

In all three FH types majority of the households indicated that women had equal access to land. However the mode of access reflected the divergent views expressed in the focus group discussions in that all FH categories, households indicated access through marriage, inheritance and both marriage and inheritance (Table 5.4).

5.2.3 Financial capital

Households in the study area have limited access to financial capital in terms of money saved either formally in banks, savings cooperatives or informally at home, with relatives, or with friends. The majority of the households have no access to credit due to lack of collateral. Investment in shares, stocks of food stored from the previous harvest for sale is also limited. Significant differences were observed among the three FH types (Table 5.2). Households in FH type two have more access to credit and indicated they have saved money in a bank or SACCO than those either in type one or two (Table 5.5). This is probably because this group has the highest proportion of educated and employed household heads who have access to bank and SACCO credit because they have collateral in terms of their jobs. They also have the highest proportion of households owning company shares which could be used as collateral for loans. Investment in shares ranged from agricultural, housing and land-buying companies to cooperatives and banking institutions.

Credit services existed in the area in the early 1970s and 1980s, provided by the Agricultural Finance Co-operation (AFC). However, this institution has long stopped providing credit since mismanagement led to its collapse in the 1990s. Although the current government is trying to revive AFC, for farmers to access loans, lending organizations require collateral. Use of land as collateral was considered very risky in the area because of previous experiences of defaulters with credit institutions. One case that was commonly quoted involved a pineapple project introduced in the area in the 1970s by Kenya Canners (now Cirio Delmonte Kenya Company). The company gave farmers credit in form of planting material to produce pineapples as out-growers for their processing plant in Thika. However, according to the farmers the credit was not comprehensive as it did not provide for all the inputs required and, coupled with poor rainfall during that time, the crop did not perform well. The farmers were unable to repay the loans which resulted to their land being auctioned for the company to recover their money. The farmers said that although many

Table 5.5. Frequencies for various financial capital variables by farming household type in Maragua.

Variable	FH type one		FH type two		FH type three	
	N	%	N	%	N	%
Saving						
Yes	21	18.9	88	93.6	7	84.8
No	90	81.1	6	6.4	39	15.2
Total	111	100	94	100	46	100
Access to credit						
Yes	14	12.6	20	25.5	7	15.2
No	97	87.4	70	74.5	39	84.8
Total	111	100	94	100	46	100
Source of credit						
Trader/shopkeeper	3	21.4	-	-	-	-
Relative/kin/in-laws	4	28.6	5	20.8	2	28.6
Informal saving or lending group	4	28.6	8	33.3	3	42.9
Bank or SACCO	1	7.1	7	29.2	2	28.6
Friend	2	14.3	4	16.7	-	-
Total	14	100	24	100	7	100
Credit use						
Buy farm assets	4	28.6	3	12.5	-	-
Off-farm business	3	21.4	6	25	2	28.6
Social obligations	4	28.6	5	20.8	3	28.6
Medical expenses	2	14.3	3	12.5	-	-
Education	1	7.1	7	29.2	2	42.9
Total	14	100	24	100	7	100

Source: Household survey, 2004.

microfinance institutions are operating in the area they were not very keen on dealing with farmers as they consider farming a risky enterprise that requires long grace periods for repayment. They prefer dealing with small business enterprises where the returns are faster and less risky.

The main credit providers for the majority of the households are traders, relatives, informal saving groups and friends (Table 5.5). Women are more likely to borrow from informal sources than men. The latter borrow from the bank or cooperatives, probably because men possess collateral, in form of land title deeds or salary, as required by the bank. Household in FH type two and three rarely take credit to buy farm inputs but to pay for education, invest in off-farm businesses and for social obligations like buying food for ceremonies. On the other hand households in FH type one rarely take credit for education but do so for all the other activities (Table 5.5).

Men tend to use credit to invest in off-farm businesses and in buying farm assets while women use it more often to pay for medical expenses. Generally the amounts borrowed are small, but range from KSh 350 to Ksh 100,000 (US$ 4.7-1,333).

5.2.4 Physical capital

At the household level, physical capital comprises farm equipment, housing, livestock and other personal household properties that can be converted into cash. The survey elicited detailed information on ownership of different household assets over time. Assets included livestock, personal household items and farm equipment.

Livestock

Livestock is important in improving the livelihoods of households in Kenya. Livestock plays a multiple role in the household, providing protein (in form of milk and meat) and manure, and it serves as security against contingencies and as display of wealth status (Moll, 2005). Focus group discussion pointed out that traditionally sheep and goats were common in every Kikuyu household but cows were mainly owned by the wealthy households for then and even today 'for a man to be called rich he must own a number of cattle'. A livestock value was calculated by multiplying the number of cows, goats and chicken, (the most common kinds of livestock in the households) with the estimated sales value of each, which was given by the households. Farming household type one has the lowest livestock value while FH type three has the highest value (Table 5.1). Although different types of livestock (chicken, ducks, turkeys, geese cows, goats, sheep and pigs) are found in the area, cows contribute significantly to the livestock value: KSh 41,700 (US$ 588) for cows compared with KSh 14,050 (US$ 198) and KSh 2,640(US$ 37) for goats and chickens respectively. The mean number of cows owned by households is two although there are households without any, especially in FH type one, while households in FH type three may have as many as 24 cows.

Cow milk production is a major income-generating activity, although dairy goats are becoming common too for household milk production. The vast majority of cattle are improved breeds, managed in zero-grazing or semi-zero-grazing systems due to lack of pasture land as a result of the small farm sizes. Cows are often fed planted fodder *Pennisetum purpureum*, (Elephant grass, also called Napier grass), banana stems, maize stover, weeds and grass, or compounded dairy feed. An important element of this farming system is the use of the manure to fertilize crops.

Generally female-headed households have fewer livestock value than male-headed ones. This is because although both male- and female-headed households have all the three types of livestock, the latter have no cows.

Farm equipment value

Households in FH type one have significantly less farm equipment than FH type two and three households (Table 5.1). The farm equipment asset value was computed by calculating the value of various farm equipment owned by the households. These include: spray pumps, water tanks, irrigation equipment, poultry houses, piggery houses, zero-grazing units, granaries, power saws, and wheel barrows. These are items that can be used to generate cash or if need be can also be sold. The most expensive farm equipment held in the households is a water tank followed by a granary, borehole and a zero-grazing unit for cattle.

Personal household items value

Again households in FH type one have significantly lower household items value compared with two and three while farm type two households have the highest. The items included in this category are a car, bicycle, television, mobile phone and a solar panel. The cars contribute the highest value although owned by only 4 percent of all the households. The majority (63%) of the households own one or two bicycles and 19 percent own one or two mobile phones. Five households own commercial buildings in the trading centres of which the resale value varied between KSh 50,000-3,000,000 (US$ 704-42,254).

5.2.5 Social capital

Using the definition of social capital as described in Chapter 2, we looked at household membership and participation in community organisations. No significant difference was observed among the three FH types in their membership in community organisations. The majority of the households (88%) have members belonging to a community organisation. Generally households in FH type two and three had higher number of household members engaged in groups than type one. The services and benefits accrued to households by virtue of their participation in these organisations were also investigated. Although most of these services are not directly accessed, being a member in the group enables households to get access to various services such as health, education, informal credit, agricultural technology and inputs that would otherwise be inaccessible. Although members are drawn from the same village, most of the community organisations in the area are heterogeneous; they differ with regard to religion, gender, age, occupation and level of education level Membership is generally voluntary, apart from a few cases where members were either born into the group (as in clan groups) and therefore are obliged to become members.

Types of community organisation, membership and involvement

Households were asked to mention two community organisations that members are involved in. The groups mentioned were: farming groups, rotating savings-and-credit associations (ROSCAs), burial societies, village committees, clan-based groups, religious groups and HIV/AIDS-associated groups. The groups were ranked by assigning weights of one and two in declining order such that rank one has the highest weight and vice versa. The overall score for each group is then calculated by summing up the number of households that mentioned the group multiplied by the rank position. This is presented by the equation below:

$$GR_1 = \sum_{i=1}^{2} Freq(1_i) \times (2 - i).$$

Where

GR_1 = Overall score for group 1

i = Rank position (1, 2)

$Freq(1_i)$ = Number of times group 1 was mentioned in rank position i

Farmer groups were mentioned as the number one most important community organization (Table 5.6). Farmer groups are organized around commodity crops and livestock and involve both production and marketing. Some of the active ones include avocado marketing, French beans production and marketing, banana production and marketing, and dairy goat rearing. In in-depth interviews with key informants, it was found that many of these were initially started as ROSCAs, but the need to generate more income had necessitated diversification into other activities. In-depth interviews revealed gender and age divisions in these groups where the youth are involved in French bean production and marketing, women and men in banana marketing, while avocado marketing groups are mainly composed of men. It was noted that there are some elderly men who are not much interested in other commodity crop groups as they had been involved in coffee production during the coffee boom and who are still hopeful that coffee production will be revived.

ROSCAs are the second important organization. According to Bouman (1995) numerous mutual aid groups with a financial component have been reported

Table 5.6. Ranking of community organizations that household heads engage in.

Type of group	Group one		Group two		Total score	Overall rank
	Freq.	Score	Freq.	Score		
Farmer group	50	100	6	6	106	1
ROSCAs	39	78	21	21	99	2
Burial Society	31	62	19	19	81	3
Neighbourhood/village committee	27	54	9	9	63	4
Clan/family	10	20	7	7	27	5
Trader or business association	2	4	2	2	8	6
Religious group	2	4	1	1	5	7
Health Group (HIV/AIDS)	1	2	1	1	3	8

Source: Household survey, 2004.

in Africa. The Ministry of Agriculture (MOA, 2004) reports that there are 154 women groups in Maragua district whose key objective is savings and lending. The functioning and organisation of these groups was investigated through in-depth interviews with members. ROSCAs are mainly composed of women. They contribute an agreed sum of money to a pool on a daily, weekly, bi-weekly or monthly basis. The pooled amount is awarded to one member of the group at a time, either in an agreed order or by drawing lots. ROSCAs vary considerably in their functioning and organization. Typical variations include membership and amount and frequency of contributions. Membership varies by age, gender and occupation. However, one fundamental requirement for membership is willingness to participate in group activities according to a set of rules and regulations. The amount to be contributed varies with the number of participating members, and the total winning amount that each member can get. Contributions can also be in the form of shares that allow a member to have more than one share, thus increasing the chances of winning the lump sum. The frequency with which contributions have to be made in each cycle also varies depending on the amount to be contributed. Usually the smaller the amount, the shorter the cycle is. The winner of the lump sum is decided by balloting or by common agreement between members, in which case the amount is usually given to a person who needs it most.

New types of savings and credit groups locally known as *Ngumbato*, literally meaning 'grasping', have sprung up in the area. These groups are organized in a slightly different way from ROSCAs and are more formal. They have a larger and more heterogeneous membership that cuts across class, economic and gender differences. The groups are initiated by microfinance institutions who introduce the idea in villages where social cohesion is thought to be high. They are established and managed by rural village committees selected by the members. One such institution that was found to be popular with most groups in the area is the Polite Service Enterprises (PSE). As opposed to village banks where the sponsoring microfinance institution provides seed money for starting up, PSE only provides training while all the start-up money is provided by the community. Polite Service Enterprise keeps records of payment and contributions for every member and provides training in book-keeping records for the officials and group members. The groups give one percent of the total revolving fund as payment for this service. Every member also pays a fee for registration (KSh 150.00) upon which they get a pass book. The group sets a minimum monthly amount of saving for every member. Members are allowed to borrow twice the amount saved or 'shares' as they call them, and for each loan they pay an interest of ten percent from the principal which is deducted from the amount given. Members co-guarantee one another, and trust and social pressure prevent default. Savings are also built up through fines for lateness and absenteeism.

Burial societies locally known as *withike wi muoyo*, which literally means 'bury yourself while still alive', are also common associations in the area. These are relatively new associations that were initiated in the area as a result of the high incidence of death in most households. They are mostly clan-based and are composed of both men and women who contribute a specified amount of money on a monthly basis towards a fund used to cater for funeral expenses. In case of a death in the family, members of the group are expected to assist in provision of food during the mourning period which lasts for about three days, as well as provide financial assistance for transportation and mortuary fees.

Village or neighbourhood associations are important in the area as well. They are generally welfare committees involved in various activities, like security, water provision, and education and caring for orphans and the elderly. This last category has become an important issue in the area because of the increased number of AIDS orphans and elderly people taking care of orphans. These community initiatives have risen out of humanitarian concern combined with reciprocity.

Membership in organisations for female-headed households was slightly higher than in male-headed households although the difference is not statistically significant. It was interesting to note that men are increasingly getting involved in group activities, a phenomenon previously predominated by women. This might be explained by the fact that men had began to appreciate the financial role groups play, particularly the ROSCAs. Previously men could easily access money through their coffee crop sales but with the decline of coffee prices they have found themselves without a regular source of income. Men are mostly involved in the formal group lending associations (*Ngumbato*), farmer groups and burial associations. Women are involved in ROSCAs and village committees (Table 5.7).

Table 5.7. Gender differences in household participation in community organisations in Maragua.

Type of group	Household heads				Wives	
	Male (N=136)		Females (N=23)			
	N	%	N	%	N	%
Farmer group	47	34.6	3	13	13	7.9
Neighbourhood/village committee	21	15.4	6	26.1	44	26.7
ROSCA	26	19.1	11	47.8	91	55.2
Health Group (HIV/AIDS)	0	0	1	4.3	1	0.6
Trade or business group	2	1.5	0	0	-	-
Burial Society	29	21.3	1	4.3	13	7.9
Clan/family group	10	7.4	0	0	2	1.2
Religious group	1	0.7	1	4.3	1	0.6
Total	136	100	23	100	165	100.0

Source: Household survey, 2004.

5.3 Household livelihood strategies and activities

Livelihood assets represent a potential and a set of possibilities for the household to secure a livelihood. The extent to which the potential of these assets is realized depends upon the way they are utilized. In the next section we are going to discuss the livelihood strategies and activities that households engage in based on their asset endowments.

Livelihood activities comprise the various activities that people do on a day-to-day basis to make a living. There is an increasing recognition that households' livelihoods in the developing world are based around a wide range of activities that seek to maximize the use of the bundle of resources and assets accessible to them (Niehof, 2004). In the study area three broad clusters of livelihood strategies are identified as: farming, livelihood diversification that includes both paid employment and rural enterprises, and migration for income generation and getting remittances.

5.3.1 Farming

Farming is the main livelihood activity of households in the study area. Most households utilize all the available land which is intensively farmed. Mixed farming involving crop and livestock production is practiced by most of the households. Crop production is a primary activity for all households; while livestock production (especially of dairy cows) is undertaken by wealthy households (see preceding section). In focus group discussions farmers ranked the five major important crops mainly based on varied criteria such as consumption, marketing, nutrition etc. During the survey households were also requested to mention what they considered their five most important crops in terms of marketing. The crops were then ranked by assigning weights from one to five in declining order such that rank one has the highest weight of five and rank five has the lowest weight of one. The overall score for each crop is then calculated by summing up the number of households that mentioned the crop multiplied by the rank position assigned to the crop as shown below:

$$CR = \sum_{i=1}^{5} Freq(A_i) \times (6 - i).$$

Where

CR_A = The overall score for crop A

i = Rank position (1, 2 ...5)

$Freq(A_{i})$ = Number of times crop A was mentioned in rank position i

The overall ranking based on farmers' perceptions was different for maize and beans from that obtained by use of frequencies. In focus group discussions farmers ranked maize as the second most important crop while in the household survey beans were ranked second (Table 5.8). This is because even though the quantity of maize produced is higher, the market price and consequently the value of beans is higher than that of maize (Table 5.9). Households may rank maize higher than beans because it is the staple food crop in the area. However, a word of caution is necessary as these results may not be representative of the actual situation because when the study was being conducted a serious drought had occurred in the area during the previous two seasons and most households had not harvested any crops. This could have resulted to the low numbers of those who indicated that they had sold surplus produce. All the crops except banana are primarily grown for own consumption with the surplus being marketed.

Table 5.8. Crop ranking based on farmers' perceptions and the number of households that are marketing the crops.

Crop ranking based on farmers' perception	Score					Total score	Overall rank
	1	2	3	4	5		
Banana	157	9	63	15	8	1,048	1
Maize	82	144	22	4	0	690	3
Bean	0	79	146	20	1	795	2
Irish potato	0	3	14	140	43	117	4
Sweet potato	0	1	0	9	40	81	5

Source: Household survey, 2004.

All across the FH types, households have diversified into other income-earning crops. Households are growing crops such as French beans, kale, (*Brassica oleracea* var. *acephala*), tomato, avocado, mango, macadamia, papaya and taro (*Colocasia esculenta* (L.) Schott). The vegetables (French beans, tomatoes and kales) are grown in the river valley bottoms using irrigation and sold in the local urban markets or in Nairobi, while the French beans are exported to Europe. Vegetable growing is usually practiced by unemployed young men as it is a labour-intensive activity that involves watering, frequent weeding, pest and disease control and frequent picking (especially the French beans). A few households (24, 16 in FH type two and three, respectively) had not cut down their coffee bushes and those who had maintained them well are currently reaping benefits from the revived coffee prices. Fruit trees are planted on hedges or randomly in the farm. Fertilizers and hybrid seeds are used for maize while manure is commonly used for bananas, potatoes and sweet potatoes. Intercropping is common for maize and beans.

Households were asked to indicate the amount of land they had cultivated, the quantity they had produced, consumed and marketed in the previous season. Again this exercise was complicated by the fact that most households had not harvested any produce in the previous season and the information given was mainly based on the previous year's activities. This presented problems in recalling the exact amounts produced, consumed and sold. The results are presented on Table 5.9. The ratio of marketed produce was computed by getting the mean of quantity produced minus the quantity consumed and given out divided by the quantity produced. The value of produce was computed by multiplying the quantity produced by the prevailing market prices at the time of the study.

Table 5.9. Differences in average land size, value and marketed ratio of the crops produced in three farm types in Maragua.

Variable	FH type one (N=111)		FH type two (N=94)		FH type three (N=46)		All categories (N=251)	
	Mean	SD	Mean	SD	Mean	SD	Mean	SD
Land under banana (ha)	0.45***	0.32	1.27	1.61	1.15	0.93	0.89	0.08
Land under bean (ha)	0.53	0.85	0.79	0.84	0.99**	0.73	0.72	0.84
Land under maize (ha)	0.67	1.8	0.96	0.97	1.04	0.62	0.84	1.35
Value of banana produced (KSh'000)	17.1	53.0	43.9	83.6	134.8*	677.7	48.3	297.3
Value of bean produced (KSh'000)	12	33	32	117	32	89.6	23.3	84.5
Value of maize produced (KSh'000)	2.9	6.1	6.9*	12.6	4.1	9.3	4.6	9.3
Ratio of marketed banana produce	0.37*	0.65	0.56	0.40	0.53	0.35	0.47	0.52
Ratio of marketed bean produce	0.19	0.36	0.21	0.34	0.19	0.39	0.20	0.35
Ratio of marketed maize produce	0.16	0.32	0.29	0.39	0.29	0.39	0.28	0.37

*, **, *** identifies variables that are significantly different at the 10%, 5% and 1% level respectively when the mean of variable is compared to the mean of the other two combined.
Source: Household survey, 2004.

Comparison between the farm types for the first three important crops shows that there is a significant difference between the categories in the land size allocated to production of banana and bean, but not maize. Households in FH type one have significantly smaller banana farms than those in FH type two and three while FH type three has the largest land size under bean production. The value of banana produced was significantly higher in FH type three than in one and two and the ratio of banana output marketed was lowest in FH type one.

5.3.2 Livelihood diversification activities

The majority (68.1%) of the households have diversified from farming into a range of off-farm income-earning activities. These include formal and informal employment, provision of services, income from renting houses and land plus remittances from migrants. Table 5.10 shows that there is significant difference between the categories in employment dependency ratio, formal and informal employment, renting and migration.

Employment dependency ratio

Employment dependency ratio is defined as the number of people formally employed compared to the number of people not working. The ratio is commonly used to indicate the degree of economic capacity of an area. Employment dependency ratio was computed for the sample population by dividing the percentage number

Table 5.10. Means and standard deviations (SD) of livelihood diversification activities for three farm types in Maragua.

Variable	FH type one (N=111)		FH type two (N=94)		FH type three (N=46)		All categories (N=251)	
	Mean	SD	Mean	SD	Mean	SD	Mean	SD
Employment dependency ratio	30,929	29,881	13,727***	23,961	32,745	38,838	24,819	30,860
Formal employment (yes=1)	0.13	0.33	0.36***	0.48	0.26	0.44	0.24	0.43
Non-formal employment (yes=1)	0.31*	0.46	0.34*	0.48	0.13	0.34	0.29	0.45
Trading (yes=1)	0.24	0.43	0.29	0.48	0.24	0.43	0.36	0.45
Service (yes=1)	0.21	0.41	0.29	0.46	0.22	0.42	0.24	0.43
Renting (yes=1)	0.1	0.1	0.09	0.28	0.09**	0.28	0.05	0.22
Migration (yes=1)	0.27**	0.45	0.44	0.50	0.30	0.47	0.34	0.47

*, **, *** identifies variables that are significantly different at the 10%, 5% and 1% level respectively when the mean of variable is compared to the mean of the other two combined.
Source: Household survey, 2004.

of household adult members who are in formal employment by the percentage of those who are not employed and multiplying by 100 as below:

(PEA / PUA) × 100.

Where
EA = Employed adults and UA = Unemployed adults

The figure for the average employment dependency ratio shown in Table 5.10 is unusually high because the zero values (representing households that had no members in formal employment) were corrected to 0.001. Notwithstanding, the average employment dependency ratio for the sample population is high which reflects the low economic activity in the area. Households in FH type two have a lower employment dependency ratio compared to those in FH type one and three. This is primarily because FH type two has the largest number of employed household heads.

Formal employment in our context refers to legally taxed employment. Most of the employed people work in the local government offices, municipal council, schools, hospital and local NGOs or hold jobs in urban centers and commute to their work place either daily or weekly. When we consider both informal and formal employment more than half of the total households have an employed person (Table 5.11). This is accounted for by the large number of people in informal employment which includes local agricultural labour, non-agricultural labour and migration for non-agricultural labour. Agricultural employment is provided by neighbouring farms that were bought from a large sisal farm (Samary) which was sub-divided and sold to individuals in the early 1980s. These are usually large farms (> 5 ha) and are mostly owned by non-residents referred to as 'telephone farmers' as they do not live on the farms. They grow labour-intensive horticultural crops for export and local market (e.g. Everest Farm which grows French beans and Bahati Farm that grows bananas, passion fruit and other horticultural crops) and rely on labour from the small-scale farms. A limited number of household members work in the hotels, hair salons, and shops in the Maragua trading centre.

Trading

About 30 percent of the households are involved in a trading activity. Farming household type two has the largest proportion of households engaged in trading activities although not significantly different from the other two categories (Table 5.11). This is probably because they have more income which they can use to start-up trading activities. Trading activities include buying and selling of maize and beans, bananas, livestock, natural products such as honey and poles. Although banana is

Table 5.11. Frequencies for livelihood activities of households by farm household type in Maragua.

Variable	FHH type one (N=111)		FHH type two (N=94)		FHH type three (N=46)		All Categories (N=254)	
	N	%	N	%	N	%	N	%
Formal employment								
Yes	14	12.6	60	63.8	12	26.1	61	24
No	97	87.7	34	36.2	34	73.9	193	76
Non-formal employment								
Yes	34	30.6	32	34	6	13	72	28.3
No	77	69.4	62	66	40	87	182	71.7
Trading								
Yes	23	20.7	34	63.8	11	23.9	74	29.1
no	88	79.3	60	36.2	35	76.1	180	70.9
Service provision								
Yes	23	20.7	27	28.7	10	21.7	62	24.4
no	88	79.3	67	71.3	36	78.3	192	75.6
Renting								
Yes	110	99.1	8	8.5	4	8.7	13	5.1
No	1	0.9	86	91.5	42	91.3	241	94.9
Migration								
Yes	30	27	41	43.6	14	30.4	86	33.9
No	81	73	53	56.4	32	69.6	168	66.1

Source: Household survey, 2004.

the main income earner for the households in the area, only 16 of the households (6.3%) are engaged in buying and selling of banana. Buying and selling of milk is also a common activity for both women and men. A dairy, owned by Kenya Creameries Cooperative (KCC), exists in Maragua town where milk is collected and delivered to a major collection and cooling centre in the district (Githumu dairy). Payments are made through *Murata* SACCO. However, delayed payments have disillusioned farmers who now prefer to sell their milk from the farm to individuals who buy and sell in urban centres without processing or packaging.

Brewing and selling local alcoholic beverages is prevalent in the area although none of the survey respondents admitted that they were involved in the activity, as it is an illegal activity. However, focus group discussions and key informants indicated that the activity is widespread in the area and associated it with increased crime, rapid spread of HIV/AIDS, and increase of irresponsible drunkard husbands. It was reported that brewing is mainly done by women while men are involved in selling the product.

Services

Services such as water, transport and health are usually provided by the government but when the government fails to provide these essential services, the community members fill in the gap. There were no significant differences between the farm types in this livelihood activity. However, again FH type two has the largest proportion of households engaged in service offering activities (Table 5.11). The activities include: ferrying domestic water from water sources (rivers and wells) to town by bicycles and donkey carts (the town is not supplied with piped water); ferrying luggage using handcarts (*mkokoteni*) and passengers with bicycle taxis (popularly known as *boda boda*). Bicycle taxis have recently become a popular income-earning activity, especially for young men who ferry people from the trading centre into the rural areas that are not well-served by public transport. These activities require some cash to purchase the bicycles and the handcarts which again explains why households in FH type two who have more people in employment and probably more income are engaged in these activities. Another activity that has gained popularity as an income earner in the area is practicing traditional medicine. Focus group discussions indicated that this is due to increased number of new diseases (HIV/AIDS-related and diabetes), the community's belief in traditional medicine, and the high cost of conventional drugs. A registered group of herbalists, known as *HEGA* Herbalists, comprising of 20 members was found practising. They are also engaged in creating HIV/AIDS awareness through teaching in churches and other forums. Other activities include digging pit latrines and hairdressing.

Renting

Renting out land, residential houses and commercial buildings is also a source of livelihood in a few households. Of the 13 households involved in this activity eight were in farm type two and only one in farm type one (Table 5.11). This activity includes renting of land located in river valley bottoms for production of vegetables and renting of commercial buildings and rental houses in the trading centres.

Crafts and small industry

A few households (5.5%) are involved in making and selling handicraft. Focus group discussions indicated that banana fibre products have recently become a popular product for both the local and the export market. Various foreign entrepreneurs are interested in promoting the products abroad and have initiated farmer groups that are making the products. One such group has been initiated by a Spanish organization working with the Maragua Comprehensive Care Centre (CCC). The group is composed of PLWA that have tested HIV positive and put on ARVs. Members of this group, who are mainly women, said they had turned to this activity because it was not strenuous and was generating some income for them.

Pottery is also an important income earning activity in Maragua; the area is famous for production of clay pots that are exported to many areas in Central Province and beyond. The pots are made locally with special clay that is unique in this area. Pottery is practiced by women who jointly hire production facilities in one of the trading centre (Gakoigo). They buy the raw materials, mainly clay, from men who dig from the deposits located in people's land. Women make and sell the pots to distant places such as the Rift Valley where they exchange them with maize and beans during the harvest periods. Other income earning activities include carpentry, house construction, metal works and dress-making.

Migration

Slightly more than one-third (34%) of the households indicated that a member had left the area during the previous year. Households in FH type two have significantly more migrated members than those in FH type one, but the difference is not significant between FH type three and two households Table (5.11). Household migrants comprise sons and daughters who are usually the young school leavers not keen on farming. Since there are no job opportunities in the rural areas they go to the urban centres of Thika, Murang'a, Nairobi and as far out as Mombasa, Eldoret and Nakuru in search of jobs. They go to these areas mainly as unskilled workers in factories and industries and to carry out informal trading activities. Young girls also move to towns for domestic labour where they are employed as living-in housemaids in urban households. According to participants in focus group discussions, Maragua town has also become an important destination for local migrants especially the youth. This was attributed to good infrastructure and services, like a good tarmac road connecting the town to Nairobi and other urban centres, a railway line with both cargo and commuter services, electricity, good telecommunication and even internet services. As one elderly man observed:

> *'The number of youth has increased in the villages and in the evening they come to the Maragua town centre. Some have even moved and rented houses in the town. The young people come to Maragua town in the evening because they have nothing to do and there is 'life' in the evening in form of video shows, drinks, drugs and disco shows. The girls mix with older men who are looking for 'Tucunguas' (directly translated 'small oranges', a*

term used by one of the local singers to refer to young girls). The men have money and entice the young girls into prostitution. In the morning they move back to the villages to work during the day.'

The majority of the migrants (53%) are involved in the so-called circular migration where they are away for between one and three months. This is because most of them are employed on a casual basis and return home once the contract is over. Although the traditional migration pattern is from rural to urban areas, due to land pressure rural to rural migration is also practiced. According to one focus group participant sub-division of land has resulted in 'small pieces of land that cannot support production even of those bananas that we are talking about'. People are migrating to other areas where land pressure and prices are low. They sell part of their land (not all because of cultural attachments) and use the same money to buy bigger units of land in other areas of the Rift Valley and Eastern Kenya (Embu and Meru).

International migration is not common in the area though popular in the country. This could be due to the fact that overseas migration is expensive and only wealthy families can afford to participate. Among the interviewed households one has two members who have moved out of the country and indeed this household is in FH type three, which as we have seen can be classified as wealthy. In-migration was reported in a few households (21). People (mostly sons and daughters) go back home from urban centres and other parts of Maragua district for various reasons. These include loss of jobs, illness, school completion, to help on the farm, marriage or as orphans when their parents die from AIDS or other causes.

About 34 percent of the households are receiving remittances from migrants. Farming household type two has a higher proportion of households that receive remittances compared with type one and three although this difference is not statistically significant. This is probably because these households have more members who have migrated. The remittances are mainly in form of food and cash.

5.4 Discussion

Classification of farm households into various farm types based on their asset endowment has brought out differences that would other wise be masked if the sample households were to be discussed all together. The data demonstrates differences in asset and resource endowments and consequent livelihood activities and strategies among the households in the three farm types. Social capital as measured by household membership in formal and informal community associations, is high across all the categories. Below we give a description of each farm type based on the assets they possess and the activities and strategies they undertake.

5.4.1 Farming household type one (N=111)

From the foregoing discussion households in FH type one category have a high proportion of female-headed households of which the heads are divorced, separated and widowed. The majority of these women are widowed. They also have a high proportion of elderly (65+) household heads and, consequently, a high demographic dependency ratio. The majority of the household heads have primary school education but the group also has the highest proportion of heads that have no formal education at all. They have small pieces of land inherited from their parents, but half of the households have no title deeds for the land. A few rent-in land to grow food crops and vegetables for sale. Most of the households have no access to formal credit presumably because they lack collateral in form of title deeds and have no savings. In terms of physical assets, these households have the least livestock, farm equipment and household items value. Thus FH type one, which form the largest category, is largely composed of poor households. Various studies have shown that female-headed households have limited access to productive resources and are poorer than male-headed households (Doss, 2001; Spring, 2000). This could be one factor that contributes to low resource endowment in FH type one households. In terms of livelihood activities households in FH type one engage in subsistence farming, where producing mainly for own consumption. Production in these farms is poor one because of the small land size and because of little or non-use of fertilisers and manure, since they lack cash to purchase these inputs and also have few animals from which they can generate manure. Some of the households have diversified into other vegetable crops for the market, such as French beans. Diversification into other livelihood activities is also low in farm type one households. They are primarily engaged in informal employment which largely involves hiring out their labour to other farms. Households in FH type one exhibit a typical subsistence orientation.

5.4.2 Farming household type two (N=94)

This group is composed almost entirely of male-headed households who are married but there are a few widows/widowers. Household heads in this category have a relatively high level of education and this group has the largest proportion of college and university graduates. There are also more households in this group that have members both in formal and non-formal employment. They have comparatively larger pieces of land (on average >0.8 ha) with title deeds. These households have some access to credit in banks and SACCO presumably because they have title deeds and savings from their employment. They also get loans from non-formal lenders because their involvement in non-formal organisations is higher than that of farm type one households. They keep more animals than FH type one households; they have more farm equipment and personal household items. Farm equipment that contributes to the high value in these households mainly includes cars, water tanks and boreholes. Of the 11 households that own cars in the study sample eight are in FH type two. With this kind of resource endowments FH type two households engage in crop production for household consumption as well as for the market. They have a higher proportion of land under banana production and market slightly more than half of their produce. They have also diversified into trading, service provision and receive remittances from migrants. Thus, FH type two households can generally be described as semi-subsistence as they are producing both for the market and home consumption.

5.4.3 Farming household type three (N=46)

This is the group with the lowest number of households (n = 46) and is composed largely of male-headed households who are married with a few widows/widowers. More than half of the households have secondary level education but their employment status does not differ much from households in FH type one. This category has the largest proportion of households with large land size (> 0.8 hectares) which is registered with title deeds. They have access to credit although few have savings or own company shares. Majority borrow from informal saving groups. They have the highest livestock value and equally high farm and personal household item value, but not as high as FH type two households. Households in this group engage in production of banana for the market and maize and beans for home consumption. They produce more beans than all the other households but most of this is consumed in the household. Most of the households in this group are not in any form of employment. However, they have the largest proportion of migrants and receive more remittances than FH type one and two households.

In the subsequent chapters the three farm typologies are referred to as low-, medium- and high- resource endowment farming households. Table 5.12 shows a summarised profile of the three farm typologies.

Table 5.12. A summarized profile of the farm typologies.

Farming household type	Summarised profile
Low-resource endowment (N=111)	Large proportion of female heads, high demographic dependency ratio, low education level, small inherited farms, without title deeds, involved in subsistence farming, and earning off-farm income by hiring out own labour.
Medium-resource endowment (N=94)	Mainly male-headed, relatively high level of education, lowest employment dependency ratio, have access to both informal and formal credit, relatively large farms partly inherited and purchased with title deeds, involved in semi-subsistence farming, earning extra income through off-farm activities and remittances.
High-resource endowment (N=46)	Male-headed, high level of education, have large farms partly inherited and purchased with title deeds and are engaged in semi-subsistence farming. Have a high physical capital in terms of livestock, farm equipment and household items. Earn extra income from remittances and rarely involved in off-farm income activities can be described as relatively wealthy, producing both for own consumption and the market.

Chapter 6

Effects of HIV/AIDS on farming household livelihood

The classification of farming households into relatively homogenous groups in the previous chapter brought out differences between the groups in terms of their asset endowments and consequent livelihood activities. However, comparison among the three farming household types using a chi-square test indicated no significant differences in HIV/AIDS incidence (Table 6.1). This conforms to the widely accepted view that unlike other sicknesses, HIV/AIDS though partly poverty-related, is not just affecting the poor. This is probably due to the fact that it is primarily spread through sexual contact thus pervading all social economic classes. Indeed some studies have found a positive correlation between socio-economic status and HIV infection (Ainsworth & Semali, 1998). However, poverty also forces some individuals to adopt risky sexual behaviour which may increase their susceptibility to HIV infection. (Chapoto & Jayne, 2005).

Consequently, the discussion on the effects of HIV/AIDS on household assets and livelihood activities in this chapter is not disaggregated by farming household type. Alternatively, households are disaggregated on the basis of their HIV/AIDS status as affected or non-affected as defined in Chapter 3. The discussion in this chapter is presented in four sections. In the first section, the differences in asset endowment between HIV/AIDS-affected and non-affected households are presented.

Table 6.1. Frequencies of affected and non-affected households in low, medium and high resource endowment households in Maragua.

Farming household type	Household HIV/AIDS status					
	Affected (N=74)		Non-affected (N=177)		Total (N=251)	
	N	%	N	%	N	%
Low-resource endowment	33	29.7	78	70.3	111	100
Medium-resource endowment	28	29.7	66	70.2	94	100
High-resource endowment	13	28.3	33	71.7	46	100

Chi-square= 0.04.
Source: Household Survey, 2004.

The second section discusses livelihood activities undertaken in HIV/AIDS-affected and non-affected households based on their asset ownership. In the third section, factors that determine the significant particular livelihood options and activities undertaken in HIV/AIDS-affected households are investigated. Finally three case studies are presented which elaborate the impacts of HIV/AIDS on various household livelihoods.

6.1 Differences in asset of HIV/AIDS-affected and non-affected farming households

Literature indicates that the levels of human, natural, financial, physical and social capital of farming households are affected by HIV/AIDS. An analysis of variance (ANOVA) was performed to investigate the differences in asset and resource ownership in HIV/AIDS-affected and non-affected farming households. Table 6.2 shows there are significant differences in various aspects of human, natural, financial capitals between affected and non-affected farming households. These are described in the following section.

6.1.1 Human capital

The most immediate impact of HIV/AIDS is on the human capital base primarily through the availability and allocation of labour. Household size is usually used as a measure of household labour. In addition, the impacts differ depending on the demographic characteristics of the household such as age, education level, and gender of the household head.

Household size

Affected households have a higher (6.6) mean household size than the non-affected households (5.3). This is contrary to what would be expected, given that affected households had lost either one or more than two members due to AIDS over the last five years. There is mixed empirical evidence on the effects of household size due to the death of a family member from AIDS. Households either dissolve or become bigger. For example, Urassa *et al.* (2001) reported that rural households in Tanzania are altogether dissolved upon the death of a household head, but Yamano & Jayne (2004) found an increase in size of household that had lost a male adult in Kenya. This was a result of the entry or return of other members. Haddad & Gillespie (2001) suggest that the new members are brought into households that have lost a family member to compensate for the loss in adult family labour. In our study the relatively large size of affected households can be attributed to the larger proportion of new members who join the household (11.8%) compared with those that left (5.3%) after the death of an AIDS members. However, the new members may not be a source of labour as they are mostly orphaned children whose parents had died of AIDS.

Table 6.2. Means and standard deviations of household characteristics for affected and non-affected households in Maragua.

Household characteristics and assets	HIV/AIDS status of the household					
	Affected (N=75)		Non-affected (N=179)		Total (N=254)	
	Mean	SD	Mean	SD	Mean	SD
Human capital						
Age of household head (years)	61.9	14.8	49.1	13.7	52.9***	15.1
Household size (no.)	6.6	2.76	5.3	2.2	5.7**	2.44
Sex of household head (1=male)	0.77	0.42	0.9	0.29	0.87**	0.33
Education level[1]	2.3	0.81	2.6	0.73	2.47*	0.76
Dependency ratio	94.2	127.5	65.9	6.1	74.3*	86.9
Natural capital						
Farm size (ha)	3.41	5	1.87	1.77	2.33***	3.17
Land/labour ratio	0.62	0.59	0.49	0.47	0.53	0.51
Land ownership[2]	1.27	0.5	1.54	0.45	1.40***	0.5
Financial capital						
Access to credit (1=yes)	0.27	0.45	0.15	0.35	0.18*	0.39
Savings (1=yes)	0.57	0.5	0.41	0.49	0.46*	0.5
Physical capital						
Livestock value	57.3	50.9	59.11	50.6	58.6	50.6
Farm equipment value	53.7	111.6	37.2	0.75	42.1	90.6
Personal household item value	31.3	131.1	18.2	72	22.1	93.3
Social capital						
Group membership (1=yes)	0.81	0.39	0.87	0.34	0.85	0.36

*, **, *** identifies variables that are significantly different at the 10%, 5% and 1% level respectively.

[1] education levels 1 = none, 2 = primary, 3 = secondary, 4 = college/university.

[2] land ownership categories 1 = own with title deed, 2 = own without title deed.

Source: Household survey, 2004.

Age

As the results in Table 6.2 show, there is significant difference of age of household heads between HIV/AIDS-affected and non-affected households. Affected households have older household heads than non-affected households. The proportion of household heads aged 60 years and above was more than twice (58.7%) that of heads in non-affected households (24.9%). In households that had experienced a death, 44 percent of the deceased were sons/daughters and only four percent reported the death of a household head. In-depth interviews revealed that most of the affected households are headed by grandparents who are fostering orphans left behind by their daughters or sons that have died from AIDS. As a result, the demographic dependency ratio in HIV/AIDS-affected households is significantly higher than in non-affected households. Previous studies (Menon *et al.*, 1998; World Bank, 1997) have not reported a significant effect of AIDS on the dependency ratio. Menon *et al.* (1998) argued that although an AIDS-related death may not necessarily result in changes in the dependency ratio, the longer-term effects are likely to be significant, particularly as the productive generation dies and the elderly are left to take care of young people. This could be the case in the sample population where the impacts of AIDS are probably starting to become manifest. In this study it was found that in households that reported a death due to AIDS 83.8 percent of the diseased members were in ages between 15 and 65. This has important implications for agriculture as premature death does not allow for transfer of farming knowledge to the younger and older generations who are now faced with challenges of farming. It may also influence technology adoption as studies have shown that age of the farmer is related to adoption decisions. Younger farmers have been found to be more knowledgeable about new practices and may be more willing to bear risk (CIMMYT, 1993). We shall return to this discussion in Chapter 7.

Education

Results indicate that the level of education of the heads in HIV/AIDS-affected households is significantly lower than that of non-affected household heads (Table 6.2). Literature in the earlier days of the epidemic postulates that individuals with higher education level were more vulnerable to HIV/AIDS (World Bank, 1999). It was argued that men with higher education and consequently higher income were more likely to travel outside their working areas and have a wide range of sexual partners. However, emerging evidence suggests that educated individuals are able to understand the consequences of having multiple sexual partners and are also more aware of ways of preventing infection (Yamano & Jayne, 2004; Blanc, 2000). This reversal is linked to increased awareness in methods of prevention and behavioural change programs and campaigns. However, we must bear in mind that majority of heads in HIV/AIDS-affected households in our sample are grandmothers whose level of education is usually low in any case. The study did not also take into account the level of education of the diseased or ill persons, which would have given a better indication of the relationship between HIV/AIDS and education of household members.

Gender

Results show that HIV/AIDS-affected households are more likely to be female-headed (Table 6.2). This further confirms our findings in discussion groups where it was constantly mentioned that grandmothers are fostering orphans given that majority of these female household heads are over 60 years old. The process through which households became female-headed was identified through in-depth interviews and comparing the proportion of single, widowed, separated and divorced women. More than half (63.6%) of the female household-heads were widows, 21.2 percent were married while the rest (15.2%) were single, separated or divorced women. The proportion of widowed women in affected households was however higher than in non-affected households. Moreover, in households that reported the death of a member due to AIDS, 51.8 percent of these members were females. These results may not be conclusive because of the small sample size and a larger sample would probably give different results. Affected households are therefore not only disadvantaged by the impact of HIV/AIDS but also have to deal with drawbacks associated with female headship, such as limited access to productive resources such as land, credit and extension services.

Land/labour ratio

The land/labour ratio, measured as the ratio of the area operated to the number of family members engaged in farming on a full-time basis, is used as an indicator of the household labour. Contrary to the expectation that HIV/AIDS-affected households would have a high land/labour ratio due to the loss of labour associated with morbidity and mortality, no significant difference was observed between affected and non-affected households (Table 6.2). This is despite the fact that affected households have a higher number of ill adult working members and experience multiple deaths of adult working members, which means that labour loss is generally higher than non-affected households (Table 6.3). The fact that land/labour ratio is not significantly different in affected and non-affected households could be due to the fact that affected households have been reported to 'import' labour which involves bringing back relatives or other members of the household previously living elsewhere. (Yamano & Jayne, 2004; Gillespie *et al.*, 2001; White & Robinson, 2000). The coping strategies employed by households in the study area to overcome labour constraints are discussed in section 6.2.1 of this chapter.

Table 6.3. Number of ill and deceased working adults in HIV/AIDS affected and non-affected households in Maragua.

Variable	Affected		Non-affected	
	N	%	N	%
No. of ill adult working members				
0	25	33.3	158	88.3
1	38	50.7	20	11.2
≥2	12	16	1	0.6
Total	75	100	179	100
No. of deceased adult working members				
0				
1	42	56	174	97.2
≥2	23	30.7	5	2.8
Total	10	13.3	0	0
	75	100	179	10

Source: Household survey, 2004.

6.1.2 Natural capital

Land size

Affected households in the sample population have significantly more land than non-affected households (Table 6.2). As land is usually used as a proxy for wealth the results imply that most of the HIV/AIDS-affected households might not be amongst the poorest. However, a significant number of the affected households have no title deeds to their land, which has implications for decisions about land use (Table 6.2). Various authors have suggested that title registration of land has several benefits, including investment in land productivity, reduction in land conflicts, easing access to credit and development of a land market (Miceli *et al.*, 2001; Smucker, 2002). This is particularly important in the context of HIV/AIDS where widows and orphans are reported to lose land after the death of the husband or parents (FAO, 2003). This is especially so when the women are suspected to be HIV-positive or to have infected their deceased husbands. Discussions in focus groups indicated that although there were land conflicts, cases of widows being sent away were rare. However, in-depth interviews revealed how some widows are struggling to continue using the land once their spouses have died (See Section 6.3).

Land transactions

Land sale in the sample population is rare as only three households had sold land in the last five years. This is contrary to what is reported in various studies which indicate that households experiencing labour loss due to HIV/AIDS dispose of their land (Du Guerny, 2002; Topouzis & Du Guerny, 1999; Gillespie, 1989). Respondents said that in the early days of the epidemic, land used to be sold to raise money to treat family members with AIDS when people thought that the disease was curable. This was because doctors were then reluctant to reveal the cause of sickness forcing the family members to move from one hospital to another looking for a cure and in the process incurring high costs. But today people know that AIDS is a terminal disease without a known cure. Indeed one of the participants expressed the general feeling prevailing in the community by saying: 'Why sell land because of a person who is going to die anyway? If for instance three people contracted the disease from the same family how much land would there be to sell?' Coupled with this is the strong attachment to land among the Kikuyu who rarely sell ancestral land. This is probably because land is seen as a source of security for the whole community and distress sale of land is typically the final act of a household on the verge of destitution. Sale of land can therefore render the household non-sustainable. Aliber & Walker (2006) reported poor households in Thika, Embu and Bondo districts of Kenya indicated that they would never sell land as it represented their last source of livelihood. Another hindrance to land sale is the requirement by the Land Control Board for consensus from the family members before any land transactions are made. This mainly because majority of the farming households do not have actual transferable rights to the land they are farming as is the case in the study area where over 50 percent of the households have no title deeds to their land. Aliber & Walker (2006) comparing the impacts of HIV/AIDS on land sale in Tanzania (Kagera), South Africa (Kwa Zulu-Natal) and Kenya reported lower incidences in the latter. They attributed the lower incident of land sale in Kenya to the significant role played by the Land Control Boards in preventing unapproved land sales.

6.1.3 Financial capital

The literature suggests that the financial base for affected households is usually low due to the need to pay for medical bills of ill persons and funeral expenses for the dead (Haddad and Gillespie, 2001; Topouzis, 2000). In this study, HIV/AIDS-affected households have more savings and access to credit than non-affected households (Table 6.2). This is despite the significantly ($P < 0.001$) high mean costs incurred for both hospitalization and funerals during the illness and after the death of affected members in HIV/AIDS-affected (KSh 27,073 (US$ 381) and KSh 5,104 (US$ 72) HIV/AIDS-affected and non-affected respectively). The majority of households (71.4 and 73.3% of HIV/AIDS-affected and non-affected respectively) used savings as the first option of financing medical and funeral costs in both households. The use of household savings is probably the first option because the loss of savings does not have an immediate adverse impact on the current welfare of the household although future household investment will be negatively impacted. The second source of finance is assistance from relatives or extended family members living away from home in form of remittances.

In 20 percent of the HIV/AIDS-affected households that had taken credit, the money was used for medical expenses, compared with 3.8 percent in non-affected households. This is probably because by the time they are experiencing the second and third illness or death, affected households have already exhausted their savings. The sources of credit for both types of households are mainly informal lending groups, savings and credit co-operatives and relatives. The average amount borrowed by HIV/AIDS-affected households is relatively small, KSh 2,314 (US$ 33), compared with KSh 3,387 (US$ 48) in non-affected households.

Apart from the direct financial costs incurred in HIV/AIDS-affected households, indirect costs are also incurred through the foregone income of the deceased and the loss of income for the care provider. In our sample population, the proportion of deceased household members that had been in formal employment is higher in HIV/AIDS-affected than in non-affected households although the difference is not significant. This means that the financial contribution of these members to the household is no longer available and this has implications on the household's ability to maintain a sustainable livelihood.

6.1.4 Physical capital

Several authors propose that HIV/AIDS-affected households have few physical assets mainly because they sell productive assets such as livestock, farm equipment and household tangible goods to raise cash to pay for health care, food, school fees, funeral expenses and other basic household needs (FAO, 2003; Ekaas, 2003; Topouzis, 2000). In this study it was found that livestock, farm equipment and personal household items values do not differ significantly between affected and non-affected households (Table 6.2).

6.1.5 Social capital

Membership in community organizations does not differ significantly between HIV/AIDS-affected and non-affected households (Table 6.2). This is not unexpected because virtually all the households in the study sample belong to a community organisation. The social capital of households operating through their relationships with extended kin and the community is said to be critical to their ability to recover from the effects of illness or death of a household member due to HIV/AIDS (Mutangadura *et al.*, 1999; Stokes, 2003). According to Mutangadura *et al.* (1999) households with little social capital to draw from are highly vulnerable to HIV/AIDS.

Table 6.4 shows the type of organization household members are involved in by gender of the household head. A high proportion of heads in HIV/AIDS-non-affected households is involved in farmer groups and burial organizations, while the majority of heads in affected households are involved in finance, credit or saving groups. The proportion of women involved in a savings-and-credit group was even higher among women in affected households.

Table 6.4. Frequencies of involvement in community organizations of affected and non-affected households by gender in Maragua.

Type of organisation	Household head				Wife	
	Affected (N=75)		Non-affected (N=179)		Affected (N=75)	Non-affected (N=179)
	Male N (%)	Female N (%)	Male N (%)	Female N (%)	N (%)	N (%)
Farmer group	9 (23.1)	2 (16.7)	38 (38.0)	1 (9.1)	2 (5.7)	11 (8.5)
Neighbourhood/village committee	7 (17.9)	4 (33.3)	14 (14)	2 (18.2)	6 (17.1)	38 (29.2)
Savings-and-credit group	8 (20.5)	5 (41.7)	20 (20.0)	6 (54.5)	25 (33.3)	66 (50.8)
Health group (HIV/AIDS)	-	1 (8.3)	-	-	-	1 (0.8)
Trader or business association	1 (2.6)	-	1 (1.0)	-	-	-
Burial Society	7 (17.9)	-	23 (23.0)	1 (9.1)	1 (2.9)	12 (9.2)
Clan/family	6 (15.4)	-	4 (4.0)	-	-	2 (9.2)
Religious group	1 (2.6)	-	-	1 (9.1)	1 (2.9)	-
Total	39	12	100	11	35	130

Source: Household survey, 2004.

Social capital may become severely strained over the long term given the systemic nature of HIV/AIDS and its clustering within families and communities (Barnett & Whiteside, 2002). To investigate this possibility in the study sample, households were requested to specify if there had been an increase, or a decrease in their participation in groups or it had remained the same over a period of five years. Table 6.5 shows that a higher proportion of affected households have reduced the number of groups they participate in, citing lack of money for membership fee. At the same time, a higher proportion of affected household had joined more groups than non-affected households. The majority (60%) of these had increased their participation in groups for the opportunity to access cash for basic household needs through the informal credit provided by the groups.

Table 6.5. Changes in group participation in affected and non-affected households in Maragua.

Household changes in group participation	Household HIV/AIDS status				
	Affected		Non-affected		$\chi2$
	N	%	N	%	
Type of change					10.8**
More no. of groups	15	25.4	23	15	
Same no. of groups	18	30.5	85	55.6	
Fewer no. of groups	26	44.1	45	29.4	
Total	59	100	153	100	
Reasons for increase					2.4
Access to cash for basic needs	9	60	10	45.5	
Increased social cohesion	4	26.7	11	50	
Agricultural produce marketing opportunity	2	13.3	1	4.5	
Total	15	100	22	100	
Reasons for decrease					7.9
Lack membership fee	12	50	32	72.7	
Poor group management	6	25	10	22.7	
Lack of awareness	2	8.3	0	0	
Lack of time	1	4.2	1	2.3	
Illness/sickness	3	12.5	1	2.3	
Total	24	100	44	100	

** indicates significance at 5% level.
Source: Household survey, 2004.

The same social relations that enhance the ease and efficiency of economic exchanges among community members have also been reported to have other, less desirable consequences. Portes (1998: 15) identifies four negative consequences of social capital: exclusion of outsiders, excess claim on group members, restrictions on individual freedom, and downward levelling norms. In our study we looked at the exclusion aspect of negative social capital and found that a significantly higher percentage of HIV/AIDS-affected than non-affected households (49.3 and 33.5% respectively) indicated that there were some community activities in which they were unable to participate. The three activities commonly mentioned include credit-and-saving groups, neighbourhood and village committees, and the local electricity project.

Focus group discussions, key informant and in-depth interviews revealed that extended family members remain the most important source of support for affected families. This is demonstrated by the fact that most affected households are fostering orphans whose parents they had also taken care of during their illness. However, the willingness and ability to support orphans among family members seemed to be weakening as demonstrated in the study cases narrated at the end of this chapter. This has prompted the evolvement of several community initiatives in the last five years. These include NGOs, Community Based Organisations (CBOs), and the 'People Living with HIV/AIDS' organization. These groups are involved in creation of HIV/AIDS awareness, providing food clothing and school fees, to orphaned

Box 6.1. Mwintindia group

This group was formed in 2002 comprising sex workers based in Maragua town who worked in the local bars and the surrounding towns of Thika, and Murang'a. It originally had 40 members but some have since died from AIDS. The group was formed with the objective of HIV/AIDS awareness creation, distribution of condoms especially in bars and lodgings, and training and counseling on management of AIDS for those who had been infected. The group was assisted by Medicos Del Mundo, Spain (doctors of the world) who are working at Maragua CCC to purchase a vehicle and facilitate their movement for awareness meetings within Maragua district. According to members the awareness campaigns have resulted in some behavioural change among members. They also believe that the training in the villages has resulted to the reported increased visits by members of the community to the VCT at Maragua CCC. Individuals are now more willing to participate in activities set up to mitigate the effects of AIDS because the awareness created by this group has helped reduce stigmatization. However, the group's activities have been hampered by the withdrawal of support from Medicos Del Mundo; now they can no longer afford to fuel the car for the training visits.

children and the elderly women fostering orphans. Some of the NGOs are offering a wide range of integrated activities, from agricultural production and marketing, skills training, micro-finance, home-care to supporting and counselling those living with AIDS. According to Gillespie *et al.* (2001) social capital at the community level is normally strengthened by the threat of a large-scale HIV/AIDS epidemic. This is usually in the form of provision of a variety of social support activities that permits families to adjust to the illness or loss of members.

6.2 Livelihood activities and strategies in HIV/AIDS-affected and non-affected farming households

The next section focuses on the differences between the livelihood activities of HIV/AIDS-affected and non-affected households and how these are influenced by farming household asset levels.

6.2.1 Farming

Several studies report that labour deficits and expenses related to HIV/AIDS morbidity and mortality result in various consequences for land use patterns (Müller, 2004; Drimie, 2002; (FAN, 2002; White & Robinson, 2000; FAO, 1995; Armstrong,, 1995). A farming household that does not have sufficient labour to farm all its land as before may choose to abandon some of its land, rent it out or sell it to other farmers to raise cash for medical, funeral or other household expenses. The household may also change from the production of labour-intensive to less labour-intensive crops. As mentioned earlier land sale is not common in the study area although of the three cases where land been sold two were HIV/AIDS-affected. In focus group discussions participants indicated that abandoning land was not common in the area among HIV/AIDS-affected households as this makes the family vulnerable to loss of land rights and grabbing by relatives and neighbours. However, household survey data revealed significant differences in land use between affected and non-affected households (Table 6.6).

A higher proportion (16%) of the affected households than non-affected households (2.8%) had abandoned part of their land due to labour shortage, lack of cash to hire labour, and lack of time due to HIV/AIDS-related illness of a household member. A higher proportion of the affected households had also rented out land to avoid bush encroachment, or raise income to meet medical expenses, payment for school fees and other basic needs for the household. Affected households are most likely to produce for home consumption as suggested by the significantly low ratio of banana output marketed. Consequently, a significantly higher proportion of affected households than non-affected reported they stopped growing labour-intensive cash crops such as French beans, tomatoes, kales, and cabbages over the last five years,

Table 6.6. Changes in household land use in HIV/AIDS affected and non-affected households in Maragua.

Coping strategy	Household HIV/AIDS status				
	Affected =75		Non-affected=179		$\chi 2$
	N	%	N	%	
Abandoned part of the land					14.7***
Yes	12	16	5	2.8	
No	63	84	175	97.2	
Total	75	100	179	100	
Rent out land					5.83*
Yes	13	17.3	13	7.3	
No	62	82.7	166	92.7	
Total	75	100	179	100	
Stopped growing some crops					21.7***
Yes	14	18.7	4	2.2	
No	61	81.3	175	97.8	
Total	75	100	179	100	

*, *** identifies variables that are significantly different at the 10% and 1% level.
Source: Household survey, 2004.

choosing to utilize the available labour on production of food crops such as maize and beans. Empirical evidence is mixed as to how HIV/AIDS is affecting cropping patterns. Yamano & Jayne (2004) found distinct shifts in cropping patterns among Kenyan households suffering a household-head death in the lower half of the income distribution. In households that incurred an adult female death there was a decrease in food crops while a male death resulted to a reduction in cash crops (tea, coffee and sugar). On the other hand, Beegle (2003) found that households experiencing a death did not shift cultivation towards subsistence food farming in Kagera, Tanzania.

Although labour shortage is expected to reduce gross production of crops produced in the farm household we found no significant difference between HIV/AIDS-affected and non-affected households in either the area planted or the amounts produced of the three major crops grown in the area. This is probably because households used various labour coping strategies. The first option in affected household for labour loss due to morbidity is to hire labour, followed by labour reallocation and, lastly, they bring relatives to live with them. Reallocation of labour is the first option used in both affected and non-affected households for labour loss due to mortality (Table 6.7). In 36 percent of the households the labour activities are re-allocated to an adult female mostly the wife.

In a few cases the activities are also reallocated to male and female children with the boys taking more of the duties than the girls. Contrary to the perception of the 'tightly knit' African community, support from village groups or relatives is rare, especially in cases of labour loss due to illness. Only 6.7 percent of the households with a sick person received any labour assistance from other households and relatives. However, assistance is higher (13.4%) in households that experienced a death. This is probably because cases of illness are not publicized especially when they are AIDS-related, whereas death is an open tragedy and people feel obliged to offer assistance. The other coping strategy mentioned is bringing in relatives. However, as already mentioned, the majority of the new members that join the households are young children and the high dependency ratio in affected households bears witness to this.

Table 6.7. Household labour coping strategies in affected and non-affected households in Maragua.

Household labour coping strategy	Household HIV/AIDS status				
	Affected		Non-affected		$\chi 2$
	N	%	N	%	
Compensation for labour loss (illness)					4.86
Hired labour	25	55.6	5	26.3	
Intra household re-allocation	17	37.8	11	57.9	
Bringing in relatives	3	6.7	3	15.8	
Total	45	100	19	100	
Compensation for labour loss (death)					0.38
Intra household re-allocation	12	60	5	71.4	
Hired labour	3	15	1	14.3	
Bringing in relatives	5	25.5	1	14.3	
Total	20	100	7	100	

Source: Household survey, 2004.

In some cases the ill person continued to work with improving condition exhibiting 'the remission-recurrence cycle' typical of AIDS (Rugalema, 1999). The remission-recurrence cycle refers to the fact that a sick person is unable to work when an illness caused by opportunistic infections is severe, but resumes normal duties when feeling better. The remission phase gets shorter as AIDS progresses. In addition no significant difference was observed between affected and non-affected households in the use of hired labour. This implies that both affected as well non-affected households compensate for labour loss by hiring-in labour. However, further probing of the participants revealed that a higher proportion of the affected households hired-in more labour to compensate both for labour lost due to illness and death than non-affected households. However, a larger sample would be required to draw definite conclusions on labour compensation.

Effects of labour loss on farming were further investigated by comparing the changes that had occurred in banana management practices between HIV/AIDS-affected and non-affected households over the last five years. A significantly higher proportion of HIV/AIDS-affected households had reduced the amount of inputs and the level of management practices in the banana plots and a significant number of affected households had altogether abandoned their plots (Table 6.8). Lack of labour, time and cash were some of reasons cited for these changes. Lack of skills and knowledge with regard to de-suckering and pruning was also mentioned in affected households.

Virtually all the households in the population sample are involved in mixed farming, growing crops as well as rearing livestock. Even though a high proportion of the affected households indicated they had sold livestock in the last five years, the livestock value did not significantly differ between affected and non-affected households. Affected households were selling cows rather than goats or chickens due to lack of labour, cows in the study area are mostly zero-grazed (i.e. fed in stalls) and therefore require substantial labour for cutting the grass from the fields and delivering it to the feeding stalls.

Extension service provision in the context of HIV/AIDS

Agricultural extension services form a fundamental part of the environment within which farming households make decisions. Although the extension services in the study area have not suffered attrition due to HIV/AIDS-related causes the staff is facing challenges in providing appropriate technology for HIV/AIDS-affected households. This is reflected in the following quote by an extension officer: 'Agriculture is labour intensive and requires people who are strong and whether you have new technologies or not it makes no difference to HIV/AIDS-affected households. What they need most is food whether the season is good or not. They

Table 6.8. Changes in banana management practices in HIV/AIDS-affected and non-affected households.

Management practice	Household HIV/AIDS status				
	Affected (N=75)		Non-affected (N=179)		χ^2
	N	%	N	%	
Reduced fertilizer usage					39.0***
Yes	25	33.3	8	4.5	
No	50	66.7	171	95.5	
Reduced manure usage					31.3***
Yes	21	28	7	3.9	
No	54	72	172	96.1	
Reduced de-suckering					19.3***
Yes	15	20	6	3.4	
No	60	80	173	96.6	
Reduced pruning					15.2***
Yes	14	18.7	7	3.9	
No	61	81.3	172	96.1	
Reduced weeding					24.0***
Yes	16	21.3	5	2.8	
No	59	78.7	174	97.2	
Abandoned banana farm					32.7
Yes	13	17.3	0	0	
No	62	82.7	179	100	

*** identifies variables that are significantly different at 1% level.
Source: Household survey, 2004.

require food throughout the year but the ministry cannot provide this. You may therefore go with your very good technology but what you find are very old women or very small children who cannot adopt anything. Ours is only the word of the mouth and these people need more than the agriculture thing we give'.

In Maragua division the number of staff is 22 and the ratio of staff to farmer is 1:1003. The division has been divided into 26 extension units and only 17 are currently covered by a extension staff. The main reason for this large ratio is the fact that the government has not been replacing after retirement and transferrals have not been replaced. This has resulted in severance of linkages and discontinuity of services in some areas. The new approach which focuses on CIGs (see Chapter 4) when handled well is one means of increasing effectiveness and efficiency of agricultural extension programmes to meet the challenges of HIV/AIDS. The CIGs act as the entry points for the extension services as well as for dealing with any cross-cutting issues such as HIV/AIDS. Since the groups are trained on fortnightly basis, other issues raised by the group such as HIV are addressed at these meetings or special training forums are organized.

In-depth interviews with the extension staff revealed that most of them have attended courses on sensitization and management of HIV/AIDS. However, these exercises have tended to be one-time events rather than an on-going process and mainly target the senior staff, leaving out the frontline staff that is directly in touch with farmers. These trainings are not followed up by activities to build upon the information imparted such as concrete initiatives to integrate HIV/AIDS into divisional work plans and budgets. It is apparent that the extension staff is facing numerous challenges in trying to disseminate information that promotes agricultural production in the current HIV/AIDS era. The staff is inadequately equipped with basic skills to deal with a clientele that is demoralized, stigmatized and has both labour and financial constraints. The extension staff is also not provided with a budget although they are expected to integrate HIV/AIDS as one of the 'cross-cutting issues' in the programme, despite the well-spelt out role of the extension service in the NAEP policy. The policy document states that 'extension services should give awareness creation maximum priority that includes passing on prevention messages including film shows on the effects of the disease and distribution of print material with special emphasis on improving nutrition standards for rural families, proper food preparation, introduction of labour-saving technologies and income generating enterprises' (Republic of Kenya, 2001c:12). Table 6.9 shows the challenges mentioned and the opportunities suggested by extension staff for themselves and the farmers in countering some of the challenges.

Table 6.9. Challenges faced by extension staff in providing extension services to HIV/AIDS-affected households in Maragua.

Challenges	How to address the challenges
Staff-related	
Lack of basic counselling and communication skills for affected households.	Training of staff on counselling and communication to affected households.
Lack of adequate knowledge on HIV/AIDS (transmission, prevention and care) Lack of knowledge on nutrition requirements for affected household members.	Training of staff on prevention, homecare, nutrition.
Inadequate financial resources to train staff and implement the knowledge learned.	HIV/AIDS should be factored in the extension programme budget.
Decision making in affected households is difficult especially where the head dies.	Reorganising extension services to target family members who are left behind such as children and the elderly.
Absent farmers and funerals disrupt the planned programmes such as field days and farm visits.	
Difficulties in reaching affected households because they withdraw from the community due to stigma.	Change attitude among the staff to reduce stigmatization.
HIV/AIDS issues are not factored in the extension budget.	Providing technical knowledge of simple activities that can be carried out by those left behind.
Increased workload for the staff due to transfers without replacement.	
Farmer-related	
Affected households are financially handicapped thus purchasing technical inputs is difficult.	Identification and promotion of income-generating projects which do not require a lot of labour and financial input.
Lack of time to implement recommendations because most of the time is used in caring for the sick.	Providing credit facilities to enable members of the surviving family to continue farming or start-up income-generating activities.
Sometimes the affected person is upset and not responsive to extension messages.	

Table 6.9. Continued.

Challenges	How to address the challenges
Farmer-related (continued)	
Households are not open about the disease due to stigmatization.	
Reduced labour, weak and sickly farmers.	
Affected families rarely think of any other issues unrelated to caring for the sick family member.	
Those who have lost partners/relatives are demoralized to continue with life (hopelessness and despair).	

Source: In-depth interviews with government extension service officers, 2004.

6.2.2 Livelihood diversification activities in HIV/AIDS-affected and non-affected farming households

Existing research on the impact of HIV/AIDS has identified livelihood diversification as a key strategy for coping with hardships such as those inflicted by HIV/AIDS (Niehof, 2004; FAO, 1995; 1998). It is argued that people need more options for securing resources and the ability to move between different livelihood strategies at times of crisis, and thereby lessen vulnerability. Among the different livelihood activities households engage in, affected and non-affected households significantly differ in renting-out of land and labour migration (Table 6.10).

To understand those factors that determine HIV/AIDS-affected households' engagement in these livelihood activities a logistic regression analysis was carried out. The following independent variables were used: age, sex and education level of the household head; dependency ratio, land size and land/labour ratio; savings, and access to credit; the personal household item value, farm equipment value and livestock value; the economic dependency ratio; and the household's membership in community group. Renting out of land and labour migration were used as the dependent dummies variables whereby 1 = yes and 0 = no. The analysis was done only on HIV/AIDS-affected households.

Table 6.10. Frequencies for livelihood activities in HIV/AIDS-affected and non-affectedhouseholds in Maragua.

Livelihood activity	HIV/AIDS status of the household						$\chi 2$
	Affected (N=75)		Non-affected (N=179)		Total (N=254)		
	N	%	N	%	N	%	
Formal employment							0.11
Yes	17	22.7	44	24.6	114	44.9	
No	58	77.3	135	75.4	140	55.1	
Informal employment							0.28
Yes	23	30.7	49	27.4	72	28.3	
No	52	69.3	130	72.6	182	71.7	
Trading							0.31
Yes	20	26.7	54	30.2	74	29.1	
No	55	73.3	125	69.8	180	70.9	
Service provision							0.55
Yes	16	21.3	46	25.7	62	24.4	
No	59	78.7	133	74.3	192	75.6	
Renting-out land							5.83*
Yes	13	17.3	13	7.3	26	10.2	
No	62	82.7	166	92.7	228	89.8	
Labour migration							11.4**
Yes	36	48	47	26.3	83	32.7	
No	39	52	132	73.7	171	67.3	

*, ** identifies variables that are significantly different at 10% and 5% level.
Source: Household survey, 2004.

Renting-out land

The age of the household head is significantly related to renting-out land in HIV/AIDS-affected households (Table 6.11). Renting-out of land is more likely to be undertaken by young household heads. In addition the gender of the household head is also significantly related to land renting but this relationship is negative. This implies that female heads in HIV/AIDS-affected households are less likely to rent-out land. Thus, despite the fact that affected households are mainly headed by elderly females they are less likely to rent out land. This is probably because

Table 6.11. Household factors influencing livelihood diversification in HIV/AIDS-affected households.

Variable	Land renting		Labour migration	
	Coefficient	S.E.	Coefficient	S.E.
Constant	7.60	3.74	-5.94	2.97
Age (years)	0.07*	0.40	0.10**	0.04
Education level	0.25	0.62	0.50	0.50
Gender (1=male)	-1.99*	1.07	-2.39**	0.89
Dependency ratio	-0.004	0.004	-0.004	0.003
Land size (ha)	-0.05	0.23	0.02	0.23
Land/labour ratio	0.10	0.86	-0.80	0.76
Hired labour (1=yes)	0.19	0.99	0.55	0.79
Saving (1=yes)	1.41	0.95	0.10	0.68
Credit (1=yes)	-0.30	0.94	0.73	0.67
Household item value (Ksh)	0.03	0.12	0.09	0.1
Livestock value (KSh)	-0.14	0.13	0.08	0.14
Farm equipment value (Ksh)	0.09	0.22	-0.05	0.15
Group membership	-0.51	0.99	1.18	0.78
Economic dependency ratio	0.75*	0.34	-0.39*	0.24

	Land renting	Labour migration
-2 log likelihood	52.1	78.0
Chi-square	16.7	24.5
Overall percentage for HIV/AIDS-affected	82.4	73.0
Overall percentage for Non-affected	92.6	69.1

*, **, identifies variables that are significantly different at 10%, 5% and level respectively.
Source: Household Survey, 2004.

they have little authority to decide on land use which is partly explained by the fact that a higher proportion of affected households have no title deeds to their land (Table 6.2). It could also be that the younger generation is not inclined to agricultural activities and opt to rent-out their land especially when they are not in any formal employment. The employment dependency ratio, which indicates the number of people in a household who are engaged in formal employment versus the unemployed, has a positive significant relationship to renting out of land in affected households. This implies that renting out of land is more likely to be undertaken in households with few employed members. Thus renting-out of land is more likely to be carried out in young male-headed households where most of the household members are not in any formal employment.

Migration

Results show that age and gender of the household head are significantly associated with migration (Table 6.11). Households headed by elderly females are more likely to have members who have migrated. The majority of the migrants in HIV/AIDS-affected households are sons, daughters or grandchildren who have moved to various urban centres in the country for employment. As expected, HIV/AIDS-affected households with a low land/labour ratio are most likely to engage in labour migration as a livelihood strategy. Although education level, land size, savings, access to credit, livestock value and being a member of a group are all not significant they are positively associated with migration. Thus, educated household heads are more likely to have contacts in urban areas through which they can acquire employment opportunities for their children. They also have more capital in terms of large land sizes, savings, access to credit, total asset value and group membership which enable them to engage in migration. Migration is an age-old strategy in the study area and recorded to have contributed more than one-half of the household incomes principally through remittances from migrants in the 1970s (Berry, 1993). However, this time-honoured activity may have now become a risk to households as it creates the channel for the flow of both cash and HIV/AIDS. A typical mechanism through which the HIV impact occurs at the household level is that of the migrant worker who falls ill while away, uses his savings for medical treatment and then returns to the farm-household to be cared for and to die, as indicated in the case studies below.

6.3 Case studies

In the next section three case studies are presented which depict the effects of HIV/AIDS on household assets and the consequent strategies adopted. The first case presents the livelihood assets and strategies of a household that is fostering an AIDS-orphan and is headed by a sickly elderly man who worries about the future of the child when he dies. The second case discusses the struggles of an AIDS widow for access to land. Finally, the third case describes the disappointment of an elderly couple fostering AIDS-orphans when they should be resting in their old age.

6.3.1 Case 1: A grandfather living with an AIDS infected granddaughter

Mwangi, a 75 year-old pensioner lives with his second wife Esther who has two children aged 18 and 15, a daughter-in-law, Lucy, and her 4 year old son, and a ten year old grand-daughter, Kanini, who is an AIDS-orphan and is HIV positive. Thus the household is comprised of seven members. Mwnagi has seven living children with his first wife, none of whom are living with him. Mwangi worked and lived in different parts of the country for many years while his first wife and children lived and worked on the farm. One time when he came home he found his wife had left, leaving a note that said that she would be back after two years. She left her eight young children who were then in school with nobody to look after them. Mwangi was devastated and even contemplated suicide. However, he soon married a barren woman to take care of his children as he continued working away from home. Unfortunately, due to interference from neighbours and extended family, this woman also left. The first wife returned home and raised her children until they had all finished school and got jobs after which she left again. Mwangi then decided to marry an uneducated woman since he thought that his first wife was arrogant and disrespectful due to her high level of education. He married his current wife Esther who is illiterate and who bore him five children. Lucy is a separated wife of one of Mwangi's sons from his first marriage who has opted to live with her father-in-law. Lucy's husband lives in Nairobi with another wife and does not come home at all. She only has primary school education. Esther's sons are still in primary school and therefore the household has no school fees burden.

Kanini is a daughter of Mwangi's deceased son Peter from his first marriage, who together with his wife died of AIDS in 2002 at the ages of 35 and 33 years respectively. They were survived by Kanini and Munene but the latter succumbed to AIDS and died nine months after his father. Kanini is HIV/AIDS-positive. After Peter and his wife passed away the wife's relatives wanted Kanini to stay with them but Mwangi resisted, arguing that she was the only connection he had with his son. So he forcibly took her away from them. However, when she came home Esther and Lucy were not willing to take care of her because they feared contracting AIDS by touching her or interacting with her. Mwangi then took her to stay with one of his sons in Murang'a, but unfortunately, when there she regularly fell sick and when she was taken to the hospital and diagnosed as HIV-positive her uncle chased her away from his home. Mwangi brought her back home where he personally nursed her and provided her with good nutrition. He also took her to the Maragua VCT where she has now been put on ARVs and goes for regular clinics. Mwangi ensures that she takes the drugs at the prescribed time without fail. He used to take her to the Maragua comprehensive health centre but she now goes on her own and is well-known by the medical staff working at the centre. He also enrolled her in the local school and informed the teachers of her status. Initially she was mistreated by the other children and taunted that her parents had died of AIDS (*mukingo*) and that she was also going to die. But the head teacher organized an awareness seminar for the pupils and teachers and this reduced stigmatization. Stigmatization has also lessened since the number of AIDS orphans in the school has increased.

At home, Mwangi has tried to reduce stigmatization by demonstrating to Esther and Lucy that they cannot become infected by sharing food or sleeping in the same room with Kanini, but they have not been very easily convinced. He has therefore taken it upon himself to take care of Kanini. They sleep in the same room and eat together, making sure she is adequately fed. However, he is worried about the girl's future as he realizes he is now old, frequently falls ill and he may not live for long. He expressed his desire to have Kanini placed in an orphan's home where he will be able to visit her when he wishes and she will also have access to proper education and care. There is no orphan home nearby apart from a feeding program run by a local community based-organization. The girl is involved in household activities such as herding. At the time of the interview (it was school holiday time) she had been sent by the step-grandmother to go and herd cattle. Peter says that Kanini expresses the wish to go to a boarding school because even before her father died he had already put her in a boarding school.

Mwangi has a daughter, Wacera, who has also been diagnosed HIV-positive and is currently sick. Wacera's husband used to work with the United Nations as a driver and passed away in 2003. One of Mwangi's sons is currently living and taking care of her and her three children. Wacera's brothers and sisters have been reluctant to reveal their sister's status but when Mwangi got to know about it he told them there is no need to hide the problem. He believes that AIDS patients should be open about it and accept their status as this has a positive psychological effect and the patient is able to find help easily once they declare their status.

Mwangi owns two hectares of land which he purchased and is in possession of a title deed. He has written a will indicating how his land will be divided among his heirs who include two wives, Kanini and any of his daughters who fail to get married. The land allocated to the first wife is currently being cultivated by one of her sons while the rest of the land is cultivated by Esther and Lucy. According to Kikuyu customs the land that is 'shown' to the mother to cultivate is generally assumed to belong to her and her children and she only has user rights. Thus Mwangi's second wife is just a custodian of her sons' land and if Mwangi passes away and her sons decide they want to divide the land among themselves it means she has no land to inherit and has to rely on her sons' generosity or good will to give her some land to cultivate. If the first wife returns home she can only be accommodated by the son who farms the land allotted to her. Mwangi believes that unmarried daughters should inherit land and has therefore allotted to Kanini, his grand-daughter, plus any other of his daughters who might never get married, the area occupied by the homestead (about 0.2 ha).

Mwangi lives in a permanent stone house with solar energy as the main source of lighting. He also owns a television set which uses solar energy. The household uses firewood for cooking and the wood is sourced from trees within the farm. He owns a big water tank in which rain water is collected for domestic use. He owns three improved dairy cows and three goats.

The pension money which Mwangi receives is not enough to cater for his household's needs because by the time he retired his salary was very little. No member of his household has any savings as far as he knows. Money was paid out for Peter as gratuity and terminal benefits (Peter was a secondary school teacher), and Mwangi has saved this in his bank account for Kanini's education. However, he laments that for him to get this money he was forced to sell three dairy cows. This is because when Peter's wife died, he had befriended a woman who after his death continued to draw his salary for three months before the employer stopped the payment. Mwangi had to pay back the money to the employer before he could get the terminal benefits. The woman has also succumbed to AIDS and her parents have contacted Mwangi to help them. He however says he is not obliged to help them since Peter had not customarily or officially married her. The hospitalisation and funeral expenses for Peter and his wife were paid by the employer. However, Mwangi paid the hospital bill incurred by his grandson who died nine months after his father.

Before Mwangi got sick he used to do a lot of farming which generated enough income for his household. Today he does not do much and most of the farm income is generated by his wife and daughter-in-law. He is not much bothered with what they do with the money as long as they feed him and Kanini. However, he complains that his wife spends all the money on buying expensive clothes. Esther had a business one time but could not sustain it due to her extravagant behaviour. According to Mwangi, his wife dresses expensively to compensate for her illiteracy. Mwangi says most people find it difficult to buy ARVs for AIDS patients because they argue that there is no need to spend money on someone who is going to die anyway. But for him he has decided that the little money he gets is spent on adequately feeding Kanini and he is ready to spend even more to buy supplements and any other foods to keep her healthy. Nevertheless, he has not sold any part of his land because according to him land cannot be sold because this would mean 'finishing' his family.

Mwangi is not member of any of the popular local community groups but he is a member of a land-buying community organization formed many years ago. This group assisted him in acquiring his land. He feels that most of the existing community organizations are mismanaged and their leaders dishonest. This perception has prevented him from becoming a member of community organizations. His wife belongs to a ROSCA that contributes money to buy household items such as plates, and cups for its members. Family ties have weakened, as exemplified by the case of his son who was not willing to take care of Kanini when she was diagnosed HIV-positive. Mwangi has a brother who is in a very well-paying profession and and although he assisted Mwangi to educate his children, today they rarely interact. He therefore is reluctant to ask him for help in taking care of his grand-daughter.

Farming is the main source of livelihood for Mwangi's household. Before he fell sick with a heart problem two years ago Mwangi used to hire labour to grow cash crops such as French beans and tomatoes. He was also keeping dairy cows, pigs and broiler chicken. Today he cannot afford to hire labour as his financial burden has been increased by his own sickness as well as that of Kanini. He has only two dairy cows now (after having sold three to cater for his late son's expenses) which provide milk for the family after he sold three cows to pay back Peter's employer (see above). Today Esther and Lucy are growing maize, beans and bananas primarily for consumption. His family has never experienced food shortage because the farm has a river valley bottom where they grow food crops during the dry season using water from the river while during the rainy season they use the upper part of the land to grow other crops when the valley is flooded. But in the previous year a certain horticultural company brought baby corn seeds to the farmers to grow in the valley bottoms and promised to collect and pay for the produce once it was ready. However, they did not collect the produce and because the farmers did not plant their usual maize crop they were seriously affected by the shortage caused by drought in the current year. The bananas are then important for food security because people could eat the crop as well as sell bananas to buy other food items required in the household. However, the income from his banana sales could only buy a small quantity of maize because the price of maize was high. Although Mwangi has heard of the new tissue-cultured bananas he has not planted them because of the high cost of the planting materials.

6.3.2 Case 2: An AIDS-widow's struggle for access to land

Life for Njoki was comfortable until her husband, Meshack, passed away in 1996 from AIDS. Meshack had been living and working in Nairobi for over 15 years and was the sole bread winner for his family. Njoki and her three children lived in the rural home where she grew crops on a subsistence level in the farm. Since Meshack passed away Njoki, who is 46 years old, says she has seen a lot of changes in their livelihood. Formerly she knew that even if she did not work so hard on the farm her husband would still provide for their needs, but now she has to work twice as hard to provide for her children. She does not expect any help from her maternal family members as they are even worse off than her. Njoki hardly ever employs labour because of the burden of school fees and does most of the work in the farm household, assisted by the children during the weekends and school holidays. When the children were younger and she wasn't paying school fees she could afford to hire labour but not anymore. She has taught her children all the skills about banana growing. Each child has her/his own stools of banana which they take care of and sell the harvest. For example during the previous Christmas holiday the eldest daughter harvested a few of her bunches (she has planted about 15 stools) and was able to buy presents for her siblings and clothes for herself.

The family lives in a permanent house which has electricity and a large water tank which stores rain water and so she does not have to fetch water from the river daily. They use firewood for cooking which is usually sourced from the farm. They have one dairy cow, a few goats and chicken. They own a television, a radio and other household items.

Meshack's parents died when he was still young and together with his four brothers they were brought up by his auntie Wanjiku. Because Wanjiku was childless she 'married' a woman to bear children for her. This woman bore five children two of whom were fathered by Meshack. When Njoki got married to Meshack, she went to live with Wanjiku while her husband lived and worked in Nairobi. However, life for her was not easy in Wanjiku's household because she was seen as competitor. She was forced to move out and join her husband in Nairobi. She stayed there for one year after which they build a house in the present land and she moved to live there in 1984. Her husband continued to live in Nairobi until 1995 when he came home complaining of chest problems. He went to Murang'a hospital where he was diagnosed as having tuberculosis. He was treated and released but had to go for injections on a regular basis. According to Njoki he did not even finish the treatment before he passed away. After this Meshack's brother Paul came to live with her but he used to drink, come home late and expect her to to let him in. Paul expected Njoki to treat him as a husband but she rejected him and he moved out and rented a house in Maragua town. Since then, Paul and Wanjiku have on several occasions tried to chase Njoki away from the home and have even threatened to kill her and her children.

Njoki and her children live on a 0.6 ha piece of land. The land is part of her father-in-law's many pieces of land which Wanjiku held in trust for Meshack and his brothers. However, Wanjiku appropriated some of the land and registered it under her name. Njoki has no title deed for her land and when she recently started following the registration in the Ministry of Lands she discovered that the land was registered under the name of Meshack's younger brother who has since passed away. This has complicated the process and for the time being she decided to leave things as they are. Paul recently filed a case with the area chief asking to be allowed to divide the land between Wanjiku, himself and Njoki. However, Njoki refused to share the land with Wanjiku saying she could only share it with Paul who would then share his portion with Wanjiku if he so chooses. Fortunately, when the case was presented before the chief and village elders, Njoki's argument carried the day and the land was divided between her and Paul. Since then Paul has on several occasions tried to sell his portion of land to strangers, but when Njoki came to know about this she requested him to sell it to her. He did and the agreement was signed in the presence of a lawyer. Although Njoki still does not have the title deed, she now uses the whole piece of land.

Njoki makes about three thousand shillings per month from the sale of bananas, milk, beans and maize. All the cash that was paid to her from the husband's final dues has long been spent. She supplies milk, maize and beans to the school that her children attend. She also fattens calves for sale to raise school fees. She has a bank account with the local post bank where she saves some money from the farm produce sales.

She has not had to sell any of her household assets although at one time her in-laws were saying that she was selling her husband's properties. But she called them for a meeting during which she asked them to prove that she had been selling assets. She told them that even if she were to sell any asset it would be those items which her husband had bought with the aim that they could be sold at one time if ever there was a need. She has restrained herself from selling any assets to avoid the rumours of the in-laws ever being confirmed or becoming true.

Njoki belongs to two community organizations which are all basically run as ROSCAs. One of the groups (Wiper Self-help group) This group which initially started as a merry-go-round wrote a proposal for AIDS awareness creation and assisting orphans which was funded by NACC. However, according to Njoki the fund (Ksh. 200,000) was misappropriated by the leaders who have even spent the money the group had initially contributed. She has abandoned this group and joined another one where each member contributes Ksh. 200 per month and members can borrow at the rate of 10%.

Today, Njoki is involved in various activities to ensure a livelihood for her family. She has intensified farming activities in various ways. When her husband was alive she had left a large part of the farm uncultivated. She has now increased the area under production by ensuring that she utilizes most of the land which had previously not been used. She also rejuvenated the banana orchards which had been destroyed by disease by replanting with the new Cavendish-resistant varieties. She has continued to plant more bananas as she realizes the cows will also benefit from feeding on the pseudo-stems and in turn produce manure for the bananas. She also sold the old cows she had and bought a better breed which produce provides enough for the family as well as surplus for sale. She sells the milk to the neighbouring boarding school where her children study. She does not receive cash payments as the school usually takes the money as part of fees for the children. Njoki has also diversified into trading. She buys maize and beans from other farmers during the harvest period which she stores and later sells when there is a shortage and during the planting season.

Njoki's household is usually not affected by food shortages because she normally buys from other people during the harvest period which she stores and later sells when there is a shortage. However, during the previous year she had to buy 90 kg of maize because her household was not categorized as needy for the food relief that was being distributed during that period. Normally households to be provided with relief food during shortage periods are identified by village elders and the chief.

6.3.3 Case 3: An elderly couple fostering AIDS orphans

Samuel is 62 and Mary his wife is 59 years old. They live with a 24 years old daughter who is unmarried and unemployed and one son who is 29 years old, unmarried and is a *matatu* (public transport vehicle) driver. They also live with three orphaned boys of a son who, together with his wife, passed away in 2002 from AIDS. The boys are aged twelve, ten and six and all are in school. The first two boys are in a local private school locally while the last one is in a boarding school that takes care of HIV/AIDS orphans in Nairobi. The first two had also been in a children's home but the grandparents removed them when they noticed they were not being well taken care of. However, the last young boy says he likes his school and appears to be healthy and happy although lately according to the grand-mother he said he missed his siblings. All the same Mary says the boy as a higher chance of getting a sponsor who will pay for his education by remaining in the school than when he is in the rural area. If he got a sponsor this would reduce their burden and they would be able to take care of the other children much better. The couple bore six children: two girls and four boys. Two of the sons are married with children and living away from home. One daughter is also married and lives elsewhere with her husband and three children. Samuel is a pensioner who worked in different parts of the country as a teacher. He retired in 1999 and came home to join his wife in her farming activities.

Samuel's eldest son, Michael, used to work in Nairobi as a public transport driver while his wife Ruth stayed in their rural home with Samuel's mother on her farm. Samuel's widowed mother allowed her grandson, Michael, to settle on this farm since Samuel had settled on a different farm and she did not have another son who could inherit the land where she was living. According to Samuel's wife Mary, her mother-in-law used to accuse Ruth of being unfaithful and said that she was moving around with men. However, since Mary loved and trusted Ruth, she never believed her mother-in-laws accusations and assumed it was the normal conflict between mother and daughter-in-law. But after some time Ruth began getting sick frequently and was diagnosed as HIV positive. She never informed her parents-in-law but when her husband got to know about it he stopped coming home and only sent food and cash. When the situation became unbearable for Ruth she went back to her maternal home and her husband took the children with him to Nairobi. However, Mary, against her husband and the other children's advice, fetched the children from Nairobi and brought them to their home. Mary feared the children would get lost in Nairobi or become parking boys. By then their mother was in and out of hospital and finally her relatives brought her back to Samuel who took her to hospital. However, Ruth was released after one week once it was realized she was terminally ill. Samuel brought Ruth to his home where Mary nursed her till she died. A month later, Michael also succumbed to HIV/AIDS-related illnesses and passed away. The couple insisted that they did not know the cause of the Michael and Ruth's illness until after their death. However, Mary says they suspected all along but 'the whole issue was very bad and unbearable'. She could have been referring to the stigma associated with AIDS. The older boy at times is lonely and sad and when asked why he says he is thinking about his parents. So even though the grand parents are providing the boys with everything they are not able to take the place of their parents.

The couple owns 1.6 ha of land which they inherited from their parents and they have a title deed for it. They live in a permanent stone house and have a borehole and a water tank from which they get water for domestic use. Sometimes they hire labour to fetch water, especially when the tank has no water during the dry periods. Generally collecting and firewood is the wife's responsibility but during the school holidays the boys assist her. They have several household items such as a television, mobile phone, and a bicycle.

The pension money together with the money from banana sales used to be adequate for the two of them since the boys came to live with them three years ago, it is not enough and they have had to increase their banana acreage. Although they have heard of the tissue cultured-banana, they have no money to buy the planting material and therefore used suckers from their old plantation for planting. Sometimes they also sell maize when the harvest is good. They sell milk but they do not get much because they only sell evening milk (about 4 litres), the morning milk is used at home. The money from milk sale is utilized to buy food items for the house. Decisions are jointly conducted on all matters i.e. what to grow, how much to plant, which inputs and how much to use etc. However, given that Samuel was working away from home Mary acquired most of the farming knowledge and skills and attended various agricultural training seminars in farmers' training centres and field days. Of late she has not been able to attend any agricultural farmers' field days due to increased responsibilities in the household. Samuel once attended an agricultural field day where they were trained on growing various crops and pest control.

They hire labour for most of the farm activities and therefore though they would have wanted to plant more bananas they do not have the cash. 'When I was young and strong I was involved in all the farming activities including marketing. I used to produce enough food for my family and have extra for sale. Now I am old and cannot do as much work' says Ruth. They have not saved money in a bank account but they own shares in a land buying company.

Currently Samuel is involved in selling milk but the cash is used by both him and Ruth to purchase what is required in the home. He delivers the milk every day to the shopping centre and is paid on a daily basis. He uses some of the money to buy house provisions while the remainder is spent on jointly agreed expenditures.

Samuel is member of an informal lending institution called Jamii Bora (Better Family). The institution lends money to men who are organized into groups of five men who know and trust one another. The group members do not necessarily have to be people from the same village. They save a minimum of KSh 50 every week which they use to open an account with the institution. After six weeks members are qualified for a loan which is double the amount they will have contributed. The group members co-guarantee one another where the group becomes the security for the loan. If any member defaults the other members are responsible for repayment. Samuel recently took credit from his group which was used by the wife to start a chicken business with one of her daughters who is a married teacher and lives away from home.

Samuel's family also belongs to a group called *Muigaruro* largely composed of people who are fostering orphans. The group is funded by NACC and is essentially involved in training although very few people go for the training sessions. At one time people used to be compensated for the time spent in these meetings and at occasions they distributed food for the orphans but since they stopped distributing food people lost interest. Mary normally attends these meetings and she thinks 'the trainings are good because I gain information and I can now tell others how to avoid catching this killer disease. If I had known or had been enlightened on HIV/AIDS, I would have helped my daughter-in-law to live for more years'.

After the death of his son and daughter-in-law, Samuel realized that the burden of taking care of the children was going to be too heavy for him and his wife and he therefore enlisted the assistance of his other children. He called them for a meeting during which each pledged to give a certain amount of money on a regular basis, especially to pay the school fees for the children. Although primary school education is free in Kenya since 2002, they did not want them to go to the public schools because initially they had taken them to boarding orphan schools in Nairobi and they realized they would not fit very well in the local schools. Mary pleaded their case with the owner of the local private school who agreed to lower the fees for the orphans. Despite the pledges made by the family members the contributions had not been consistent. According to Samuel and Mary in spite of providing good education for their children they are not holding very well-paying jobs and the little money they get is barely enough to provide for their own families. In fact, they still come back to their parents for additional support. 'Will the government ever create well-paying jobs so that our children will be able to stand on their own?' asks Mary.

> *'When at times we discuss with my husband we wonder whether a time will come when we will have no burdens because now we have gone back to bringing up another generation instead of resting and eating the fruits of our labour. The little money my husband got on retirement we spent to ensure that our children studied well but now we have to take care of their children as well. Oh God, when are we ever going to rest!'*

6.4 Discussion

It is evident from both the case studies and the survey data that HIV/AIDS is changing the size and structure of households in the study area. HIV/AIDS-affected households are mostly female-headed and have a significantly higher dependency ratio and experience a greater shortage of labour despite their larger household size. This is because they tend to take care of more, younger orphans, and thus have few family members in the economically active age category. In Case 1 and 3 the household size has increased as a result of the orphans that have joined. In Case 3 Samuel is able to hire some labour to augment farm production although labour constraint still limits the household's farming activities. In Case 1, Mwangi cannot afford hired labour due to health costs. The household size in Case 2 was reduced due to the death of the household head although the dependency ratio has increased. The household lacks cash to hire labour due to the school fees burden which implies that the workload for the woman has increased. This case portrays labour constraints in female-headed households.

Contrary to the frequently quoted strategy of selling land due to labour loss and increased household costs related to HIV/AIDS, land sale is rare in the study area. White & Robinson (2000) reported that although sale of land as a coping strategy in response to HIV/AIDS shock had been reported in literature, differences had been observed in different geographical areas. For example the FAO identified sale of land at low prices as a household coping strategy in response to HIV/AIDS (FAO, 1997). In contrast, a household survey in Kagera, Tanzania recorded no land sales as a coping strategy to mitigate the impact of HIV/AIDS death or sickness while in Thailand 41% of the households that experienced a death sold land (Armstrong, 1995). Aliber & Walker (2006) categorized threat to land tenure into 'tenure events': land disputes; distress sales of land; perceived threats to tenure which included actual attempts to grab land and actual tenure loss. They found no significant difference between HIV/AIDS-affected and non-affected households in any of the land tenure events. Farm households are understandably reluctant to sell land because this is the only primary natural capital that they have control over..

Instead, a significant number of affected households have changed land use patterns by stopping to grow cash crops such as French beans tomatoes, and kales, which are labour intensive preferring to use their limited labour to ensure food security by producing food crops such as maize and beans. The household in Case 1 has stopped growing horticultural crops because its lack cash to hire labour. Besides, affected households have reduced the amount of inputs in the banana plots while some have altogether abandoned their orchards. This has reduced banana yields to quantities just enough for own consumption with no surplus for the market. The loss of farming knowledge is imminent in Case 1 since the household head seems to be the only one in possession of appropriate agricultural management skills. In contrast, the woman in Case 2 is passing on farming skills and knowledge to her children which will ensure their survival in the event of her death.

Although affected households have more savings than non-affected households, they have high direct and indirect costs. These include direct medical and funeral costs as well as the forgone income of the deceased persons. Direct costs are mainly financed through borrowing from informal sources and transfers from relatives and extended family members and credit. Cases 1 and 3 have a regular source of income through pension payments though the money is insufficient to finance their household needs. The household in Case 2 lost its bread winner but the widow is involved in various income diversification activities to ensure her family's financial needs are met. Despite the financial needs in Case 1 the elderly man, Mwangi, has saved money in the bank for the orphans' medical treatment and education which will be used in the event of his death.

Even though there is no difference in physical capital (livestock, farm equipment and household item value) between affected and non-affected households, a greater proportion of affected households are selling livestock (mostly cows) to cater for household income needs such as school fees. Evidence from the three case studies and from focus group discussions point to the importance of livestock as a source of livelihood, and most households in the area attempt to conserve their livestock as long as possible. Apart from Case 1, where the household sold three dairy cows to meet HIV/AIDS related financial costs, the other cases have not sold any of their assets. On the contrary, the widow in Case 2 sold her low grade cows and bought a breed that produces more milk. In Case 3, the household continues to maintain the cows as they are an important source of income through milk sale.

Results from the case studies indicate that social capital through membership in community organizations is a crucial asset, as it enables households to access informal credit. However, this vital social resource is becoming increasingly unavailable to HIV/AIDS-affected households who, due to financial constraints, cannot afford to pay the mandatory membership fee and the monthly payments. Consequently, a tendency to reduce participation in community organizations was observed. The fact that households are willing to foster orphans is an indication that the social capital of households operating through their relationships with extended kin is critical for enabling them to recover from the impacts of HIV/AIDS. However, this source of support seems to be threatened by stigmatization, as demonstrated in case 1 where the son of the elderly man was not willing to live with the AIDS-infected orphan once he discovered she was HIV positive. Even in his own household stigmatization is still prevalent. Poor economic status of extended family members also contributes to weakened social support as illustrated in Case 3 where family members have been unable to honour their pledge of support to their parents. In addition social reproduction in terms of role-modelling is likely to be weakened and future generations may not experience the informal exchanges of knowledge that are often embodied in a family environment.

Probably because affected households tend to have more land but less labour they are unable to utilize all the land due to labour constraints. They are engaged in renting-out land as a major livelihood activity. However, gender and age differences are apparent in that elderly female-headed households are less likely to rent-out land probably because they do not have power to decide on how land is used. Case 2 depicts the land disputes and struggles that women go through after the death of their spouses. This is because of the traditional patriarchal tenure and inheritance systems in the area where women gain access to land through their association with men (Spring, 2000). This widow is constantly fighting with her in-laws to ensure she continues to use the land left to her after her husband's death. Fortunately she has the support of the local administration and the village elders, which enables her to continue farming the land. This finding concurs with Aliber & Walker's (2006) suggestion that the link between loss of land rights and HIV/AIDS is related to how 'AIDS affectedness' impacts a person's social capital. In this case the widow was able to garner support from the leaders and community members. However, she still has no title deed and therefore cannot easily decide on land usage or transfer the land to her children. She has recently bought her brother-in-law's piece of land and obtained legal documents for it. In this way she has solved the problem of access to land and she can easily transfer the land to her children.

Migration is another important livelihood strategy undertaken in HIV/AIDS-affected households. Ironically, this strategy renders households more vulnerable to HIV/AIDS as spouses separated for extended periods of time are more likely to have extra-marital partners, thus putting themselves at the risk of HIV infection. In addition, young people who migrate from the rural to urban areas are also vulnerable.

Although the extension staff in the study area is not directly affected by HIV/AIDS their ability to deliver appropriate services to affected farming households is hampered in several ways. They lack basic counselling and communication skills and knowledge on nutrition requirements for AIDS-infected individuals. They are also facing budget and human capacity constrain.

Both the quantitative and qualitative data have demonstrated that households respond in varied ways and not necessarily as depicted in the literature. This implies that the development interventions targeting to improve livelihoods and reduce poverty in rural agricultural settings need to take into account these differences.

Chapter 7

Significance of tissue-cultured banana for farming households' livelihoods

Increasing agricultural productivity is one of the important adaptations that farming households undertake to enable them to attain sustainable livelihoods in times of crisis. Adoption of appropriate agricultural technologies is crucial for increasing productivity and rural household income (Smith *et al.*, 1994). Yet several factors influence the adoption of agricultural technologies, which may constraint the potential of the technology in contributing to development and poverty reduction. In this chapter I examine the factors that determine the adoption of tissue-cultured banana. The first section describes how the tissue-cultured banana was introduced in the study area. The second section presents farmers' perceptions and opinions of the tissue-cultured banana and explores the reasons why farmers are adopting or not adopting the technology. The adoption is then disaggregated into low, medium and high resource endowment farming households. In the subsequent section a multivariate analysis is conducted to investigate factors influencing adoption and the continued use of tissue-cultured banana plantlets for further expansion. In the final section, the livelihood outcomes associated with adoption of tissue-cultured banana are discussed.

7.1 Tissue-cultured banana in Maragua: introduction and labour requirements

In Kenya, tissue-cultured banana technology was introduced as a package which included the provision of tissue-cultured plantlets, disease and pest tolerant varieties, and information on crop husbandry and post-harvest handling practices. The varieties introduced include seven from the Cavendish group (Chinese Dwarf, Grand Nain, Williams, Giant Cavendish, Valery and Dwarf Cavendish) and the Goldfinger variety. The Cavendish dessert varieties are tolerant to *Fusarium* wilt race 1, which is the prevailing race of fungus in the study area. Goldfinger is a variety developed at the Honduran Foundation for Agricultural Research la Lima. It is resistant to various pests and diseases and adaptable to a wide range of ecological environments. It can be taken as dessert and be cooked as a staple (Wambugu & Kiome, 2001). The management package includes the types and amounts of fertilizers, pests and disease control measures, and field management practices (de-suckering, de-leafing, male flower bud removal, etc.). The post-harvest management package includes proper harvesting techniques, post harvest treatments, packaging and ripening. The technology was introduced through collaborative research between the Kenya Agricultural Research Institute (KARI), the International Services for Acquisition of Agri-biotech Applications (ISAAA), the Institute for Tropical and Sub-tropical Crops (ITSC) of South Africa, the Rockefeller Foundation and the International Development Research Centre (IDRC) in Canada.

In the study area, demonstration trials were initially conducted to compare the conventional planting material (suckers) and the tissue-cultured plantlets. Farmers participating in these trials were provided with tissue-cultured plantlets and suckers for free. The demonstration trials were followed by a pilot micro-credit program to facilitate adoption of the technology. The programme, introduced in the area by ISAAA and KARI in collaboration with a local NGO, was conducted in a modified Grameen lending model whereby members of farmer groups were provided with tissue-cultured plantlets on credit. The group acted as the collateral for the planting material. The groups were modelled around the Farmer Field School concept, whereby during the monthly group meetings members were trained on management aspects of the banana crop throughout its complete cycle (Onyango *et al.*, 2004). The credit was to be repaid after the first harvest but members were required to make a monthly contribution of KSh 100.00 (US$ 1.5) to the group's account. About 40 percent of the interviewed households growing tissue-cultured banana were involved in this program.

7.1.1 Labour requirements of tissue-cultured banana

In his study Qaim (1999) showed that labour costs for the establishment of a banana plantation and the recurrent annual costs are about three times higher for farmers using tissue-cultured plants in Kenya. The main labour intensive activities in banana production are land preparation, planting, manure application, weeding, watering, de-suckering, de-leafing, propping, harvesting and marketing. Watering is the major labour demanding activity, demonstrated by the fact that 89 percent of the households indicated that water requirement for tissue-cultured is higher than for conventional bananas (Table 7.1). Conventionally banana growing in Maragua is rain fed and although planting is normally done at the onset of rainfall inadequate amounts usually necessitate watering. The high water demand for tissue-cultured plantlets is explained by the fact that unlike conventional suckers which have a rhizome that acts as a storage reserve for the initial growth of roots and leaves, the growth of tissue-cultured plantlets depends on their own roots and leaves thus are more influenced by external conditions. Studies have also shown that the leaves of tissue-cultured plants are photosynthetically more active than those of conventional suckers and this efficiency more than makes up for the lack of a rhizome. The fact that tissue-cultured plants have their own initial roots and continue growing as soon as they are planted plus their more active leaves means that they grow faster than conventional suckers. However, for this efficiency to be realized the plants

Table 7.1. Comparison of non-tissue cultured and tissue-cultured banana by farming households' growing tissue-cultured bananas.

Characteristic	Number of farmers expressing opinion	Non-tissue-cultured higher (%)	Tissue-cultured higher (%)	Same (%)
Labour requirement	64	4.7	76.6	18.8
Quantity of fertilizer used	60	1.7	88.3	10
Quantity of manure used	64	4.7	71.9	23.4
Water requirement	64	1.6	89.1	9.4
Pest and disease susceptibility	64	62.5	35.9	1.6
Cost of planting material	64	6.3	90.6	4.7
Yield	63	25.4	65.1	7.9
Quality	63	28.6	58.7	12.7

Source: Household survey, 2004.
NB: not all households growing tissue-cultured bananas gave a response.

must not be subjected to any external growth constraints such as lack of water or nutrients in the first five months of plant establishment (Robinson, 1996).

The other high labour cost activity is digging holes for planting. It is recommended that tissue culture plants are planted in deep holes to reduce the tendency of the mats pushing up. The pushing up of mats is attributed to vigorous growth of juvenile tissue-cultured plants due to the carry-over effect of the hormones used in the tissue culture medium (Robinson, 1996). Plants with exposed mats are poorly rooted and can easily be blown down by winds.

Another practice which was cited as being laborious is de-suckering (the removal of excess suckers). This is because tissue-cultured plants produce numerous suckers soon after planting and these have to be removed continuously to avoid competition for nutrients and water with the mother plant as this reduces the first harvest yield (Robinson, 1996).

Cultural pest control methods are also considered labour demanding. Although the use of tissue-cultured plantlets reduces infestation in new plantations and delay pest populations build up (Gold, 2001; Reddy *et al.*, 1998), pest control soon becomes necessary as the plants are frequently planted in fields that are usually infested with nematodes and banana weevils. The recommended practice is to either use pesticides or cultural control methods. The latter are preferable as they are environmentally friendly and cost-effective especially in subsistence production systems. The cultural pest control methods involve destruction of crop residue and use of pseudo-stems as traps. It is recommended that after harvesting the pseudo-stem should be cut down into small pieces which are then split open or removed from the field. Pieces of the cut stems can also be left around the banana whereby they act as traps by attracting weevils which can then be picked and destroyed. Trapping has been found to reduce weevil population by 43 percent in farmer-managed fields (Gold, 2001). As well as being laborious the application of cultural control methods requires knowledge and skills. The farming households may also weigh the trade-off in using the crop residual as animal feed and for production of farm yard manure as opposed to using it for pest control. Thus availability of labour is expected to influence the adoption of the technology.

7.2 Farmers' perceptions of tissue-cultured banana versus the conventional banana

The perceived attributes of an agricultural technology has been reported to influence adoption behaviour (Adesina & Zinnah, 1993; Mbaga-Semgalawe & Folmer, 2000). Adoption or non-adoption of technologies may reflect farmer's perception of appropriateness of the characteristics of the technology. We assessed the farmers perceptions of tissue-cultured banana by asking farming households growing tissue-cultured banana to compare the input requirements of tissue-cultured banana with the conventional planting material in terms of labour, level of inputs (fertilizers, manure, and water), cost of planting material, pests and disease susceptibility, yield and quality. The majority of households perceived tissue-cultured bananas as more labour intensive, requiring more fertilizer, manure and water but less susceptible to diseases and pests than the conventional banana (Table 7.1). Most households thought that the tissue-cultured plantlets are also more expensive than conventional planting material. However, these high input requirements seem to translate into high yields and quality of the tissue-cultured compared to the conventional banana. This is probably because the tissue-cultured banana varieties introduced are also tolerant to *Fusarium* wilt, one of the most important diseases in the study area. The varieties are also less susceptible to weevil damage than other groups of bananas (Gold & Messiaen, 2000). It could also be because most farmers who have adopted tissue-cultured bananas are growing them on land that had not been planted with bananas before, thus delaying the build up of pests and the associated consequent decline in yield. Focus group discussion with traders indicated that tissue-cultured bananas could not be distinguished in the market as such and were therefore not difficult to sell.

Despite obvious benefits, the percentage of farmers growing tissue-cultured banana in the total sample is surprisingly low, just 26 percent. Households were asked to give three main reasons as to why they were not growing tissue-cultured banana. The reasons were then ranked by assigning weights from one to three in declining order such that rank one has the highest weight of three while rank three has the lowest weight of one. The overall rank for each reason is then calculated by summing up the number of households that mentioned the reason multiplied by the rank position assigned as shown below.

Ranking calculated by the equation:

$$OR_1 = \sum_{i=1}^{5} Freq(O_i) \times (4 - i)$$

Where

OR_1 = The overall score for objective 1
i = Rank position (1, 2, 3)
$Freq(O_i)$ = Number of times reason 1 is mentioned in rank position i

Three major reasons for not growing tissue-cultured banana were mentioned. The first is lack of information on various aspects of the technology (Table 7.2). A majority of the households said they had never heard of tissue-cultured banana, a few did not know where to get the planting material from, while a few simply said they had no interest in the technology. The second reason was the higher cost of the tissue-cultured plantlets than that of the conventional suckers. Suckers are usually exchanged freely or at a minimal cost among the farming households. The third reason was the high requirement of water for the tissue-cultured banana plantlets during the field crop establishment stage.

In his *ex ante* study on the impact of tissue-cultured banana in Kenya, Qaim (1999) predicted that small-scale farmers, the lowest resource endowment group in his study, would take up tissue cultured-banana more slowly than the larger ones. This he attributed to the high initial cost of establishment mainly contributed by the cost of tissue-cultured plantlets (Table 7.3). He predicted that reducing the price of banana tissue-cultured plantlets to half of the prevailing price by then (that is from KSh 75 to 35) would increase the adoption rate from one to five after five years.

Table 7.2. Reasons for not growing tissue-cultured banana in Maragua (N=187).

Reasons	1		2		3		Total score	Overall rank
	Freq.	Score	Freq.	Score	Freq.	score		
Lack of information	74	222	5	10	9	9	241	1
High cost of tissue culture plantlets	71	213	6	12	2	2	227	2
High water requirement	26	78	45	90	0	0	168	3
High labour requirement	2	18	6	12	48	48	78	4
Scarcity of land	11	33	1	2	1	1	36	5
High manure requirement	3	9	6	12	6	6	27	6
High fertiliser requirement	0	0	10	20	8	8	18	7

Source: Household survey, 2004.

Table 7.3. Banana annual input and total labour costs with and without the use of tissue culture technology in 1998 (Ksh/ha) according to farm size.

Item	Small-scale		Medium-scale		Large-scale	
	With	Without	With	Without	With	Without
Tissue-cultured plants (1,097 plants)	82,317	6,932	82,317	7,922	82,317	10,397
Fertilisers	1,646	0	2,744	385	4,117	715
Manure	6,302	4,098	9,454	6,485	17,332	13,866
Pesticides	1,463	113	2,488	242	3,659	1,098
Total	79,728	11,143	97,003	15,034	10,7425	26,076

Source: Qaim (1999:16).

7.3 Differences among the farming household types

The classification of farming households into comparatively homogenous groups (see Chapter 5) brought out differences in the categories in terms of resource endowments. These differences are expected to influence households' responses to various interventions. A significant difference was also observed among the farming household types in the number of households growing tissue-cultured banana. In Chapter 5 the characteristics of the three farming household types are briefly described as follows: the low resource-endowment category (N = 111) that is characterised by significantly many female heads, a high demographic dependency ratio and low level of education. In this category the farms are small and inherited from the parents, though without title deeds in most cases (51%). The households are involved in subsistence farming and they earn off-farm income by hiring out their own labour. The medium resource-endowment category is composed of 94 semi-subsistence farmers who produce both for own consumption and the market. They have relatively large farm sizes partly inherited and purchased and own title deeds for them. They are mainly headed by males who have a relatively high level of education and have access to both informal and formal credit. They earn income through on-farm as well as off-farm activities such as trading, service provision, and they receive remittances. The high resource-endowment category is the smallest (N = 46). The households in this category also have large farm partly inherited and purchased with title deeds and are engaged in semi-subsistence farming. They have assets in the form of livestock, farm equipment and household items. They are not involved in off-farm activities but receive extra income through remittances.

A higher proportion of the medium and high resource-endowment farming households are growing tissue-cultured banana compared with the low-resource category (Table 7.4). Access to informal as well as formal credit, more sources of off-farm income from trading, service provision and remittances from migrants in both medium and high resources farming households probably contribute to this difference. The higher level of education of household heads probably also enables them to have better access to information about the new technology, perhaps because they are better connected to people in the research and extension system or farm inputs traders. Although the low resource-endowment group also has access to informal

Table 7.4. Distribution of households growing and not growing tissue cultured banana by farming household type.

Farming household type	Growing tissue-cultured		Not growing tissue-cultured		N
	N	%	N	%	
Low resource endowment	16	14	95	86	111
Medium resource endowment	36	38	58	62	94
High resource endowment	14	30	32	70	46

χ^2 = 15.5, p<0.001.
Source: Household survey, 2004.

credit such as friends, relatives, and ROCSAs, we saw in Chapter 5 that the amounts borrowed are small and just enough to cater for basic necessities.

Further examination of the size of land under tissue-cultured banana among growers in the three farm types shows no difference although actual sizes differ among growers (Table 7.5). The fact that the average size is equal among the growers across the farm types probably reflects that they are growing the recommended size of a feasible commercial unit of banana plot. Demonstration trials of tissue-cultured banana by researchers found that the minimum number of tissue-cultured plants required for a commercial unit is 80 plants (Mbogo *et al.*, 2002). It could also mean that land size is not a determining factor in adoption. It demonstrates that farming households are growing tissue-cultured bananas as an additional crop to conventional banana and other crops such as maize and beans.

From the foregoing discussion it is apparent that various factors influence the adoption of the tissue-cultured banana technology. In the next section we define adoption in the study context and discuss the factors that may determine the adoption of tissue-cultured banana.

Table 7.5. Average land size of tissue-cultured banana growers, land under tissue-cultured banana, conventional banana and total average land under banana in low, medium and high resource endowment farming household.

Land size	Level of resource endowment							
	Low (N=16)		Medium (N=36)		High (N=14)		Average (N=67)	
	Mean	SD	Mean	SD	Mean	SD	Mean	SD
Average land size of tissue-cultured growers	0.5	0.3	1.6	2.7	1.4	1.0	1.3	2.1
Average land size under tissue-cultured banana	0.2	0.2	0.2	0.2	0.2	0.2	0.2	0.2
Average land size under conventional banana	0.08	0.09	0.3	0.7	0.3	0.4	0.3	0.4
Average land under banana	0.2	0.1	0.5	0.7	0.5	0.5	0.4	0.6

Source: Household survey, 2004.

7.4 Adoption defined

In this study, two stages of adoption are analysed. In the first stage adoption is defined as a dichotomous decision between 1 = growing tissue-cultured banana and 0 = not growing tissue-cultured banana. A farming household is considered as an adopter if it is growing one or more tissue-cultured bananas and a non-adopter if it is not growing tissue-cultured bananas at all. Once the response on adoption or non-adoption is obtained, continued use of tissue-cultured plantlets among the adopters was investigated. A dummy variable was used where 1 = continued use of tissue-cultured plants and 0 = discontinued use. If farming households continued or intended to continue using tissue-cultured plantlets for further expansion of their banana plots they were categorised as continued users and vice versa. This was only done with households that had adopted tissue-cultured banana. It was used as a measure of adoption because there is a tendency for farmers to use suckers from tissue-cultured mother plants for expansion instead of buying new tissue-cultured plantlets. Indeed some of the participants were of the opinion that suckers from tissue-cultured plants were as good as the original mother plants. Although studies on the advantages of using first generation suckers of tissue-cultured plants in Kenya have not been conclusive, farmers are recommended to always use tissue-cultured plants when establishing new orchards or expanding their banana orchards (Wambugu & Kiome, 2001). This is because suckers are likely to be contaminated by soil borne pests entailing a risk of spreading diseases and pest, which would defeat the very purpose of using tissue-cultured plants.

Therefore we assumed two dependent variables Y_{1i} and Y_{2i} which represent farming households' tissue-cultured banana adoption and continued use of tissue-cultured plantlets.

Thus:

$$Y_{1i} = \begin{cases} 1 \text{ if household is growing tissue-cultured banana} \\ 0 \text{ if household is not growing tissue-cultured banana} \end{cases}$$

$$Y_{2i} = \begin{cases} 1 \text{ if household intention to expand using tissue-cultured plantlets} \\ 0 \text{ if household intention to expand not using tissue-cultured plantlets} \end{cases}$$

We are interested in determining the probability that a farming household adopts tissue-cultured banana $P(y_{1i})$ and continues using tissue-cultured plantlets for further expansion $P(y_{2i})$. Following Field (2000) the probability was assessed using the logistic regression analysis and is given as:

$$P(y_{1i}) = \frac{1}{1 + e^{-z}} \qquad P(y_{2i}) = \frac{1}{1 + e^{-z}}$$

Where
$Z = \beta_0 + \beta_1 X_1 + \beta_2 X_2 + ... \beta_n X_n + \varepsilon_i$

Where β_0 is a constant, $\beta_{1,} \beta_2 .... \beta_n$ are the estimated coefficients of the corresponding variables X_1, X_2 ...... X_n. The residual term is represented by ε_i.

7.5 Empirical analysis and variables

The explanatory variables used in understanding adoption behaviour of farming households in the sample population include: characteristics of the farmer, the farming household assets and off-farm livelihood activities.

7.5.1 Farmer characteristics, human capital

Age and education level and gender of the household head, household size, number of adult household members, use of hired labour, household morbidity and mortality and the health status of household members are important aspects of human capital that are associated with agricultural technology adoption decisions.

Age

The relationship between age of the household head, often used as a measure of the farmer's experience, and technology adoption can either be positive or negative. On the one hand, young household heads may be more knowledgeable about new practices and more willing to bear risk due to their longer planning horizons. On the other hand, older household heads may have more experience, resources or authority that allows them to decide positively on technology adoption (CIMMYT, 1993). In this study older household heads are expected to adopt tissue-cultured banana because of their authority to decide on land use (keeping in mind that banana is a perennial crop).

Education

Many adoption studies show some association between technology adoption and level of education of the farmer. Educated farmers with are said to be early adopters of technology and apply the associated inputs more effectively (Feder et al.1985). However this may depend on the complexity of the technology. It is argued that education may make a farmer more receptive to technical recommendations that require a certain level of literacy. In our study the level of education of the household head is expected to have positive influence on tissue-cultured banana adoption since the technology requires some level of literacy.

Gender

Women farmers play a key role in most African farming households. However, it is difficult to summarize gender roles within African households because they tend to change, particularly so with changing economic circumstances (Doss, 2001). It has been reported that although tradition often specifies some tasks or crops as women's and some as men's, these gender divisions of labour become less rigid as new opportunities arise. Although men may not take over women's responsibility of food production for home consumption they move into women's activities when these become profitable. According to Doss (2001) a new technology which increases the profitability of a crop may cause this kind of change. Banana is usually termed a women's crop in Kenya (Qaim, 1999; Wambugu and Kiome, 2001; Wambugu, 2001). However, with the introduction of tissue-cultured banana this may change and thus it is not possible to predict the effect of gender beforehand. A dummy variable for the sex of the household head was used to investigate the effect of gender on adoption of tissue-cultured banana.

Labour availability

One measure frequently used in adoption studies to account for labour availability is household size. Household size is either measured as all household members, adult household members or adult equivalents. However, no clear relationship between household size and adoption has been found. It can be positively or negatively related to adoption depending on the availability of land and labour markets (Doss *et al.*, 2002). Another variable often used is whether or not hired labour is used on the farm (Doss, 2006). In this study household size, the number of adult household members and the use and non-use of hired labour were used. As was discussed above (section 7.1) availability of labour is crucial for the adoption of tissue-cultured banana and consequently a positive relationship between adoption and these variables is expected.

In addition to these conventional measures of households' access to labour, variables to measure the quality of household labour were included in the regression. The effect of the household HIV/AIDS status was assessed with a dummy variable indicating whether a household is HIV/AIDS-affected or non-affected. The number of ill household members at the time of study is also used as a measure of quality and quantity of household labour. The number adult deaths experienced in a household is also used as an indicator of labour loss that is expected to influence adoption. The inclusion of these variables captures the influence of the vulnerability presented by HIV/AIDS to farming households' labour. In Chapter 6 we saw that households are adopting various strategies to deal with constraints associated with labour loss due to morbidity and mortality of members. However, since the effects of HIV/AIDS may be confounded by several other factors such as lack of financial, physical or social capital it is difficult to determine its effect on technology adoption *a priori*.

Access to information/extension

The uptake of new technologies is often influenced by the farmer's contact with extension services, since extension agents provide improved inputs and technical advice. An extension services variable was included to measure the access to information. A dummy variable for extension contact was created by a combination of two variables: for farmers who are growing tissue-cultured banana they were asked to indicate how they came to learn about the technology while those not growing were asked to give reasons why they were not growing tissue-cultured banana. If the reason for the latter was due to lack of awareness this was taken as an indication of absence of extension contact.

7.5.2 Farming household assets

Natural capital

Wealth is often positively associated with the adoption of new technologies, because wealthier farmers are better able to bear risk and therefore are more likely to try new technologies. In Kenya, land ownership provides a good measure of wealth. However the relationship between farm size and technology adoption is confounded by other factors such as the fixed costs of adoption, human capital, credit constraints, tenure arrangements and characteristics of the technology (Place & Swallow, 2000; Feder *et al.*, 1985). For example, large households with large farms may have more access to credit and thus may adopt a technology that has high costs. It is therefore difficult to determine the effect of land size *a priori*. Place & Swallow (2000) suggest that farmers are more likely to invest in land which they have transferability rights for example in the form of title deeds. The relationship between tenure and tissue-cultured banana adoption was represented by a dummy variable indicating ownership/no ownership of land title deed. It is expected to be positive.

Financial capital

Capital in the form of credit or savings is often required to finance the inputs associated with a new technology (Reardon *et al.*, 1995; Imai, 2003). The relationship between financial capital and adoption depends on the characteristic of the technology and the household's resource endowments such as size of land. For example, scale-neutral technologies such as new varieties (or the tissue-cultured banana) may not show a positive relationship with access to credit. Conversely, the relationship between credit and technology adoption is often confounded by land size as farm households with large farms have more access to credit. Therefore, the effect of credit is difficult to determine beforehand. To assess the effect of financial capital three dummy variables were used; access to credit and ownership of cash savings and remittances. There is no consensus in the literature about the use of remittances. On one hand there are those arguing that remittances are dedicated to land purchases and agricultural improvements. On the other hand there are those arguing that remittances are invested in consumer items and other goods and services that immediately improve the family welfare and social status. This group argues that investment in agricultural activities tend not to be the first priority. For example, Tiffen *et al.* (1994) provide an example of agricultural improvement resulting, in part, from off-farm earnings of migrants in East Africa. De Haan (1999) in a review of various studies on the effect of remittances on economic productivity, concludes that the way remittances are used depends on the form of migration, and the characteristics of the migrants and those who stay behind. In this study, we expect a positive effect of remittances on tissue-cultured banana adoption.

Physical capital

The ownership of farm equipment is reported to positively influence technology adoption (Feder *et al.*1985). In this study the important physical capital that is expected to influence adoption of tissue-cultured banana is related to farm equipment and livestock. Ownership of farm equipment such as a water tank, water pump, irrigation pipes and a borehole is important to access irrigation water, which is crucial for tissue-cultured banana establishment. Thus the relationship with farm equipment is expected to be positive. Ownership of livestock is also expected to positively influence tissue-cultured banana adoption. As well as being an important fungible asset, the sale of products such as milk provides cash to buy farm inputs. Livestock are also a critical source of manure in farming households which is used to grow bananas plus other crops.

Social capital

A dummy variable for the household's membership in a group organization is included in the regression. In practice this variable captures learning effects because extension agents largely work through local farmer organisations. It could also capture financial assets effects if the farmer's organisations double as micro-credit groups. Farmer groups are the most important community association for farming households in the study population (see Chapter 5). These pre-existing farmer groups were a major entry point in the introduction of tissue-cultured bananas in the study area. A household's membership in a group organization is expected to be positively related to growing tissue-cultured banana.

7.5.3 Off-farm livelihood activities

Participation in off-farm activities is a significant source of income for farming households in the study area. The pursuit of multiple activities can have important implications for cash and labour availability and, hence, for the adoption of the technology. The income earned from these activities may be used to purchase tissue-cultured plantlets or hire labour. By contrast, off-farm activities may compete with labour and time required in tissue-cultured banana growing. Therefore the impact of off-farm activities on tissue-cultured banana adoption is difficult to predict *a priori*. Five types of off-farm livelihood activities are included as dummy variables in the regression: trading, service provision, formal and informal employment and remittances.

The descriptive statistics of the variables used in the adoption of tissue-cultured banana and the continued use of tissue-cultured banana plantlets are shown in Annex Table A1.1 and A1.2.

7.6 Results and discussion

7.6.1 Determinants of tissue-cultured banana adoption

Several variables are significantly influencing tissue-cultured banana adoption as depicted in Table 7.6. Surprisingly, the indicators for labour availability used in this study do not have a significant relationship with adoption of tissue-cultured banana. This could indicate that at least at this stage and scale, labour availability does not affect tissue-cultured banana adoption decisions. Even more puzzling is the non-significant effect of HIV/AIDS given that tissue-cultured banana is a labour-intensive crop especially during the initial establishment stage. This is probably because, as we saw in Chapter 6, HIV/AIDS-affected households adopt various labour coping strategies such as hiring in labour, labour reallocation among household members, and bringing in new members.

The sex of the household head was not significantly related to adoption. Doss (2006) cautions that the inclusion of a dummy variable for the gender of the household head may fail to reveal the intra-household gender dynamics. We tried to unravel the intra-household labour division of labour, through focus group discussions and in-depth interviews. Gender differences in the initial response to the tissue-cultured banana technology were identified. In households that have small pieces of land women were not willing to plant the crop as they felt that this would reduce the amount of land for food crops such as maize and beans. They also felt adoption of the tissue-cultured banana would increase their labour burden because of the watering requirements. The men, on the other hand, saw the opportunity of earning

Table 7.6. Factors influencing adoption of tissue-cultured banana in Maragua.

Explanatory variable	Coefficient	Standard error	Significance level
Constant	-3.87	1.86	0.04
Age (years)	0.01	0.02	n.s
Education (level)	-0.28	0.26	n.s
Gender(1=male)	0.59	0.58	n.s
Adult household members (number between 15 and 65)	0.23	0.14	n.s
Hired labour (1=yes)	0.38	0.41	n.s
Household's HIV/AIDS status (1=affected, 0=non affected)	-0.34	0.46	n.s
Extension contact (1=yes)	0.12	0.37	n.s
Size of land (ha)	0.09	0.14	n.s
Land tenure (Ownership of title deed 1= yes)	1.00	0.46	0.03
Savings (1=yes)	0.68	0.42	0.10
Access to credit (1=yes)	0.43	0.47	n.s
Remittances (1=yes)	0.40	0.46	n.s
Livestock value (KSh)	0.01	0.08	n.s
Farm equipment value(KSh)	0.25	0.11	0.02
Personal household items value (KSh)	-0.04	0.05	n.s
Group membership (1=yes)	0.25	0.53	n.s
Formal employment (1=yes)	-1.23.	0.50	0.10
Informal employment (1=yes)	1.17	0.49	0.02
Trade (1=yes)	2.30	0.79	0.004
Service provision (1=yes)	-3.38	0.87	0.000
Land renting (1=yes)	-0.03	0.59	n.s.

Source: Household survey, 2004, N = 248.
Dependent variable = growing tissue culture banana (1 = yes, 0 = no); -2 log likelihood = 216.9; cases predicted correctly = 79.4 percent; n.s.= not significant.

more income by growing tissue-cultured banana and so took over the supervision and even management of the production. In one household where the wife refused to water the plants, the husband planted the bananas and watered the crop until it was well established in the field. One man in the group discussions said: 'If you see money is going to come you cannot just let things be done without plan', actually meaning you cannot just leave it to the women.

Land size had no significant influence. The quality of land, an aspect that was not investigated in this study, could be a more important aspect than land size in regard to the prevalence of pests and diseases in the soil. This is because although the use of clean planting material reduces infestation to new plantations and delay pest populations build up, due to small land sizes most of the tissue-cultured plants are frequently planted in fields that are infested with nematodes and banana weevils. Studies have shown that weevil damage could result in 44 percent yield loss in the third cycle of the plantation (Rukazambuga *et al.*, 1998). Thus lack of land that has not been contaminated by soil borne pests and diseases may limit the farm household from adopting tissue-cultured banana. This aspect needs further investigation.

Land tenure is positively related to adoption, as expected. Farming households with title deeds for their land are more likely to invest in tissue-cultured banana production than those without. This is probably because they have user as well as transferable rights which allow them to determine land use especially with banana which is considered a perennial crop in the study area.

Farm households with savings are more likely to grow tissue-cultured bananas. The accessibility to credit is not significantly associated with growing tissue-cultured banana. This is probably because there are no sources of formal credit for agricultural activities in the area. Even households who have access to informal credit rarely use it to purchase farm inputs.

As expected, high farm equipment value increases the probability of the adoption of tissue-cultured banana. In Chapter 5 we saw that the item that contributes the highest farm equipment value in the household is a water tank. These are usually concrete tanks that harvest rain water coming from the iron roofs of houses in the homestead. The water stored in tanks is usually used for domestic purposes but some households are also storing piped water from the Kenya Railways water supply system into these water tanks, which they use for irrigation. Thus ownership of water tanks is important given the crucial need for water during the first five months of tissue-cultured plants growth. Surprisingly, ownership of livestock is not significantly related to adoption of tissue-cultured banana although the relationship is positive.

Although involvement in community organizations was expected to link the farming households to sources of information and informal credit thus enabling the utilization of agricultural technologies, group membership does not influence tissue-cultured banana adoption. This is probably because, as it emerged from focus group discussions, most of the banana farmer groups which were used in the initial dissemination program by KARI were no longer in existence.

Engagement in trading and informal employment has a positive significant relationship with adoption of tissue-cultured banana. The most common trading activities are grain (maize and beans) and banana selling. The grains and bananas are mostly produced in the farm but households also buy from other farmers. The money raised through trading may be helping these households to invest in tissue-cultured banana production. Informal employment which largely involves agricultural labour is positively related to adoption of tissue-cultured banana. Reliance on agricultural labour is a major source of income especially for the low resource endowment farming households in the study area. The external income is important as it enables the households to undertake agricultural activities that increase their subsistence income. Thus some of the cash raised through this activity is probably being used in the purchase of banana tissue-cultured plantlets.

Formal employment and service provision are, on the contrary, negatively related to tissue-cultured banana adoption. Involvement in these off-farm activities may be competing with labour required for tissue-cultured banana farming related activities. As opposed to non-formal employment where household members work on an hourly basis and thus have time to work on their own farms, formal employment and service provision are full-time activities.

7.6.2 Determinants of continued use of tissue-cultured banana plantlets

Adoption of tissue-cultured banana alone may not lead to increased productivity unless farmers continue to use the clean planting material in their new fields. Thus the determinants of continued use of tissue-cultured banana plantlets are identified in the second step of the analysis. Continued use of tissue-cultured banana plantlets is influenced by farmer characteristics, contact with the extension service, farm assets and off-farm livelihood activities (Table 7.7).

Farming households headed by older males are more likely to continue using tissue-cultured plantlets. Large households are also more likely to continue using tissue-cultured plantlets. This is probably because they have more household labour as use of hired labour is negatively related to continued use implying that farming households that continue using tissue-cultured plantlets rely on family labour.

Households' HIV/AIDS status does not seem to influence continued use. However, the death of an adult household member negatively influences continued use. The greater the number of adult household members lost through death the less likely

Table 7.7. Factors influencing the continued use of tissue-cultured banana in Maragua.

Explanatory variable	Coefficient	Standard error	Significance level
Constant	-25.32	11.55	0.03
Age (years)	0.14	0.07	0.07
Gender(1=male)	0.66	1.93	n.s.
Household size	2.20	1.05	0.03
Morbidity (number of ill persons)	-3.21	2.23	n.s.
Mortality (number of adult deaths)	-10.58	5.23	0.05
Household's HIV/AIDS status (1=affected, 0=non-affected)	5.39	3.35	n.s
Hired labour (1=yes)	-2.63	2.07	n.s
Extension contact (1=yes)	8.72	3.89	0.03
Land tenure (ownership of title deed; 1= yes)	8.84	5.25	0.09
Savings (1=yes)	-1.08	1.82	n.s.
Access to credit (1=yes)	8.87	4.90	0.07
Remittances (1=yes)	-6.98	3.57	0.05
Farm equipment	0.03	0.24	n.s
Group membership (1=yes)	-2.45	2.06	n.s
Formal employment (1=yes)	8.35	4.23	0.05
Informal employment (1=yes)	8.87	3.93	0.02
Trade (1=yes)	2.84	1.96	n.s

Source: Household survey, 2004.
Dependent variable = continued use of tissue-cultured plantlets (1 = yes, 0 = no); -2 log likelihood = 28.9; cases predicted correctly = 92.4 percent; N = 66; n.s. = not significant.

a household is to continue using tissue-cultured banana plantlets. As we saw in Chapter 6 significantly more HIV/AIDS-affected households have lost more than one household member compared with non-affected households. In addition, a significantly higher proportion of HIV/AIDS-affected households had reduced the amount of inputs and the level of management practices in the banana plots while some households had altogether abandoned their plots. Lack of labour, time and cash were some of reasons cited for these changes. Lack of skills and knowledge in regard to de-suckering and pruning was also mentioned in affected households. This last reason could be an implication of loss of skills and knowledge associated with the death of adult members and it could also be due to lack of information due to limited extension contact. This could imply that HIVAIDS-affected households are not able to continue using the technology probably due to labour loss or the costs associated with mortality.

Farming households that have contact with extension services continued to use tissue-cultured plantlets. This is probably because they could easily get the plantlets through the extension services that have contact with laboratories and nurseries providing the tissue-cultured plantlets.

Land tenure security through ownership of a title deed influences adoption as well as continued use of tissue cultured plantlets.

Whereas access to credit has no influence in the farming households' initial decision of adoption, it is a determining factor in continued use of tissue-cultured banana plantlets. This is probably because a substantial percentage (40%) of the households growing tissue-cultured banana had been provided the initial planting material on credit. The credit was to be repaid after the first harvest of the crop. For them to continue using tissue-cultured plants for expansion they will need credit to purchase the plantlets and the associated farm inputs.

Equally, formal employment which has a negative influence on initial adoption is positively related to continued use of tissue-cultured plantlets. Farming households engaged in informal employment are also more likely to continue using tissue-cultured banana plantlets. The income earned in these activities probably enables these households to buy the tissue-cultured plantlets for further expansion.

Surprisingly, remittances had no significant relationship to tissue-cultured banana adoption while there was a negative significant influence on continued use of tissue-cultured banana plantlets. Several authors have reported that remittances from migrants do not lead to extensive changes in agriculture (De Haan, 1999; Brad, 2002). Brad argues that if remittances are sufficiently large, households may not rely on agriculture as a livelihood option because they substitute remittances for agricultural income. In the study area farming households usually receive remittances from their migrant sons and daughters which they normally use to buy food, clothing and sometimes pay school fees. Generally in other parts of rural areas in the country where people are receiving large amounts of remittances especially from international migrants, the money is usually invested in education and residential homes.

7.7 Livelihood outcomes

Livelihood outcomes are the accomplishments and gains that households derive from carrying out specific activities. Households combine their assets to pursue different livelihood activities based on their aspirations or objectives. According to Dorward (2006) people usually try to sustain their current welfare status or improve it by expanding their present activities or moving into new activities. Farming households in the study population are primarily seeking to ensure that they meet basic needs such as adequate food, clothing, shelter, education of children, maintain good health for family members and increase farm production by purchasing farm inputs and hiring labour. Thus a farming household will adopt a technology with the expectation of increased production or reduction of labour input. Increased production leads to household food security either through increased quantities for household consumption or increased income to purchase food. The income may also be invested in the farm or used to diversify the household's activities. Livelihood outputs may not necessarily be the end point as they feed back into the future asset base of the household. According to Niehof (2004:325), livelihood outcomes are thus 'part of a dynamic process, rather than the end point of a pathway going in one direction only'. Outcomes may include strengthened asset base, and improvements in well-being aspects such as food security and health.

Livelihood outcomes in this study are indicated by increased banana production, income and food availability. Households were asked to indicate whether the quantity of bananas produced, the household income and food had increased since they started growing tissue cultured banana. A chi-square test was used to assess differences in these aspects between HIV/AIDS-affected and non-affected households. An attempt was also made to assess whether a relationship existed between the outcomes and the household's asset base or improvements in well-being (e.g. health) by looking at how the extra income gained through the sale of banana was being utilised. Table 7.8 shows that a higher proportion of HIV/AIDS affected than non--affected households reported an increase in production, income and food attributed to the adoption of tissue-cultured banana. These results are based on self-reported subjective effects and not objectively measured effects. As a consequence, affected households are more likely to report positive changes, however small they may be, which non-affected households might dismiss. In both HIV/AIDS-affected and non-affected households the extra income earned by increased sale of banana is used first to ensure food security and secondly to pay for the education of the children. In doing this, the outcome feeds back into the asset base of the household by strengthening the human capital. The extra income is also used to buy clothing, household items and pay for medical bills thus contributing to the well-being of the households.

Table 7.8. Self-reported livelihood outcomes in HIV/AIDS-affected and non-affected households growing tissue cultured bananas.

Livelihood outcome	HIV/AIDS status of the household				
	Affected (N=20)		Non-affected (N=45)		χ^2
	N	%	N	%	
Increased banana production					2.47*
Yes	16	80	27	60	
No	4	20	18	40	
Increased household income					3.96*
Yes	17	85	27	60	
No	3	15	18	40	
Increased food security					3.56*
Yes	16	80	25	55.6	
No	4	20	20	44.4	

* Indicates figures that are significant at 10%.
Source: Household survey, 2004.

7.8 Discussion

The results show that savings, farm equipment, land tenure and engagement in informal employment are positively related to the adoption of tissue-cultured banana. However, formal employment and service provision are negatively related to tissue-cultured banana adoption. Continued use of tissue-cultured banana plantlets is positively related to the age of the household head and household size. The death of an adult member of the household is negatively related to tissue-cultured banana adoption. Contact with extension services, land tenure, access to credit, and engagement in formal and informal employment are positively related to tissue-cultured banana adoption. However, households receiving remittances do not seem to use the cash in agricultural production as the relationship between tissue-cultured banana adoption and remittances is negative. Notwithstanding, access to credit positively influences tissue-cultured banana adoption. Cash decreases a new technology's possibilities of adoption, as lack of cash may prevent poor farmers from investing in a technology which will not yield a return until after harvest, several months later as is the case with banana. Availability of water is also an important determining factor for the adoption of tissue-cultured banana. Having secure land rights also seems to encourage the household to invest in a perennial crop such as the banana. It is no surprise then that the high- and medium-resource endowment farming households are growing tissue-cultured banana. Literature on technology adoption is in general agreement that wealthier farmers typically adopt before poorer farmers (Kerr & Kolavalli, 1999). Wealthier farmers are capable of absorbing the risk associated with trying out the new technology, whereas poor farmers might wait to see how it performs on a neighbour's field. According to Qaim's (1999) prediction, resource-poor farmers may not be able to adopt large quantities of tissue-cultured plants.

Continued use of tissue-cultured banana is influenced by more factors. The age of the household and land tenure are important probably because older household heads have more user rights and can easily make decisions on land use. Availability of labour also becomes an important factor as demonstrated by the significant effect of household size and the death of an adult household number. Engagement in formal employment, which has a negative influence on initial adoption, becomes important for continued use of tissue-cultured plantlets probably because the cash from this activity is used for the purchase of tissue-cultured plantlets and hiring labour. Contact with extension services also encourages continued use possibly because of the linkages established for the acquisition of tissue-cultured plants.

Diversification into income generating activities is positively related to tissue-cultured banana adoption. The income earned through trading activities allows farming households to adopt tissue-cultured banana. However, engagement in service provision is negatively related to growing tissue-cultured banana probably because it engages household labour on a full time basis.

Adoption of tissue-cultured banana has implication for the gender division of labour. Though traditionally known as a woman crop, the prospects of the banana becoming a major cash crop in the study area could change this. Labour allocations may also have to be renegotiated with the adoption of tissue culture as the need for watering the plants increases women's labour burden. This reinforces the need to provide supplementary irrigation facilities to mitigate this effect. There is therefore need for development agents look into technologies that reduce women's labour burdens and provide greater control for women over the outputs of their labour.

Ownership of land title deeds by these households may also encourage a long term investment such as the banana which is normally grown for not less than five years. Place and Swallow (2000) suggest that purchased land is more likely to receive more investment as the owner has transferability rights.

Chapter 8

Conclusion and discussion

The previous chapters have focused on the identification of assets and resources, the effects of HIV/AIDS on the assets and the consequent livelihood activities and options of farming households in the study area. The study focused on banana farming as one of the farming households' livelihood activities and studied the determinants of adoption of the tissue-cultured banana and the livelihood outcomes of households that have adopted the technology. In this concluding chapter the findings of the study are first discussed in line with the research questions formulated in Chapter 1. In Section 8.2 the results are discussed in the context of the sustainable livelihood approach. Finally, policy implications of the findings are discussed in Section 8.3.

8.1 Answering the research questions

8.1.1 Assets, activities and strategies of farming households

> *Question 1: How do assets available to farming households influence their livelihood options, activities and outcomes?*

This question was answered by looking at the farming households' assets and how the assets influence the livelihood options and activities of the households. The gender relationships existing between access to assets, livelihood options and activities were also studied. Farming households were found to be heterogeneous in terms of assets endowments and it became necessary to create farm typologies to depict these differences. The households were categorised into three farm typologies: low-, medium- and high-resource endowment.

Low-resource endowment farming households

This category has a significantly high proportion of female-headed households, the majority of whom are widowed. They have a high proportion of elderly (65+) household heads and, consequently, a high demographic dependency ratio. The majority of the household heads have primary school education but the group also has the highest proportion of heads that have no formal education at all. In terms of natural capital, households in this category have small pieces of land inherited from their parents, and half of them have no title deeds for the land. Some are renting in land to grow food crops and vegetables for sale. The households have no access to formal credit and no savings. Their physical capital is also low as they have few livestock, farm equipment and few household items. They are producing mainly for own consumption and earn extra income by selling their own labour in agricultural activities.

Medium-resource endowment farming households

This group is composed almost entirely of households that are male-headed, the majority of whom have relatively high level of education ranging from primary to college or university level. They have large pieces of land (on average >0.8 ha) with title deeds. They use hired labour for farm activities. These households have some savings and access to both formal and informal credit. Farming households in this category keep more animals; they have more farm equipment and personal household items than the others. They mostly produce for own consumption and also for the market. They earn income through formal and non-formal employment as well, and are engaged in trading, service provision. They receive remittances from migrant household members.

High-resource endowment farming households

This group is composed largely of male-headed households, the majority of who have secondary education level. This category has the largest proportion of households with large land size that is registered with title deeds. They have access to informal credit but do not have savings. They have the highest livestock and farm equipment value. The majority of households in this group do not take on other income diversification activities and are mostly engaged in production of banana for the market and maize and beans for home consumption. However, they have the largest proportion of migrants and receive more remittances than the other two categories of households.

Social capital

Social capital merits particular mention here because of its major role in many of the households in the study area. In this research membership in a community organisation was taken as an indication of social capital. The majority of the households in the sample population have members belonging to a community organisation. Membership in a community organisation enables households to access services such as access to credit that would otherwise be out of reach for the majority of them. The community groups existing in the area are: farming groups, rotating savings-and-credit associations (ROSCAs), burial societies, village committees, clan-based groups, religious groups and HIV/AIDS-associated groups. Generally households in medium- and high-resource endowment categories have a higher number of household members engaged in groups than those in the low-resource endowment category. A high level of membership in community organisations is an indication of the high social capital in the study area. The Literature indicates that those communities endowed with a diverse stock of social networks and civic associations are in a stronger position to confront poverty and vulnerability, resolve disputes, and take advantage of new opportunities (Woolcock & Narayan, 2000). In their study on the role of social capital in rural Tanzania, Narayan & Pritchett (1999) found that an increase in social capital, measured by membership in community groups, was associated with higher welfare of the households in the villages. The absence of formal insurance, financial systems, well-functioning marketing systems and adequate agricultural information delivery systems in the study area has prompted the proliferation of self-help groups, burial associations and ROSCAs. However, having high levels of social capital in the form membership of informal groups does not necessarily lead to economic prosperity (Narayan & Nyamwaya, 1996). This is especially true when the groups are unconnected to outside resources as is the case in the study area.

8.1.2 The effects of HIV/AIDS on farming households' livelihood portfolio

> *Question 2: What are the impacts of HIV/AIDS on farming households' assets and consequently livelihood options, activities and outcomes?*

We first investigated the people's perception of the HIV/AIDS situation in the study area through focus group discussions and in-depth interviews. Secondly, since the incidence of HIV/AIDS on the three farm types did not differ significantly, the discussion on the effects of HIV/AIDS on household assets and livelihood activities was disaggregated on the basis of their HIV/AIDS status as affected or non-affected. The livelihood activities undertaken in HIV/AIDS-affected and non-affected households based on their asset ownership were studied. The study examined factors that determine the significant particular livelihood options and activities undertaken in HIV/AIDS-affected households. Finally the results were triangulated with information gathered through case studies.

It is apparent that most people are aware of the causes, transmission and prevention methods of HIV/AIDS. However, stigmatisation associated with AIDS is still an overriding factor that is undermining prevention in the population. Participants in focus group discussions acknowledged that AIDS-related stigma has many consequences that can affect the infected individual's life as well as others. Due to stigmatization, high-risk individuals continue to avoid testing while HIV positive individuals do not disclose their status and risk infecting others. People living with AIDS also fail to take advantage of the treatment that is available to maintain their health and well being.

HIV/AIDS-affected households are mostly female-headed and have a significantly higher dependency ratio than non-affected households. They also experience a greater shortage of labour, despite their larger household size. The relatively large size of the affected households is attributed to the orphans that have joined these households after the death of their parents. Moreover, the level of education of the heads in HIV/AIDS-affected households is significantly lower than that of heads in non-affected households.

Distress land sale due to labour loss and increased household costs related to HIV/AIDS is uncommon in the study area. Instead, a significant number of affected households have stopped growing cash crops such as French beans, tomatoes, and kales, which are all labour-intensive, preferring to use their limited labour to ensure food security by producing food crops such as maize and beans.

While affected households have more savings than non-affected households, they have high direct and indirect costs. Direct costs are met by borrowing from informal sources and transfers from relatives and friends. Indirect costs incurred through the loss of income when a bread winner passes away, cause other members of the household to engage in multiple livelihood activities to ensure livelihood security. Affected households are more likely to be involved in renting-out land and labour migration as major livelihood activities than non-affected households. However, gender and age differences are apparent in that elderly female-headed households are less likely to rent-out land probably because they do not have power in land-use decisions.

Although quantitative data shows no difference between affected and non-affected households in group membership, the case studies indicate the importance of group membership for affected households as it enables them to access informal credit. In addition, other forms of social capital in the form of kin and family networks emerge in the case studies. But differences are also apparent in these networks in that parents are more willing than siblings to foster or support orphans. This could be due to the low economic status of extended family members who also have to ensure their own livelihood security.

8.1.3 The significance of tissue-cultured banana for farming households' livelihoods

Question 3: What is the role of the tissue-cultured banana in the livelihood activities and outcomes of farming households?

The question was addressed by assessing household's perception of the tissue-cultured banana compared to the conventional planting material in relation to their assets. The reasons why farmers are adopting or not adopting the technology were investigated. The differences in adoption between the low-, medium- and high-resource endowment farming households and the factors that determine the adoption of tissue-cultured banana were studied. The factors that determine continued use of tissue-cultured banana plantlets for further expansion were also explored. We also looked at how livelihood outcomes are shaped by the adoption of tissue-cultured banana.

The majority of farming households growing tissue-cultured banana indicated that the technology has high input requirements in terms of planting material, fertilizer, manure, water, and labour. More than half of the households indicated that tissue-cultured plants give higher yield and better quality bananas than conventional plants, indicating a willingness to adopt the technology. Surprisingly, despite obvious benefits, the percentage of farmers growing tissue-cultured banana in the total sample was low (26%). Farmers mentioned lack of information on various aspects of the technology, the high cost and high requirement of water for the tissue-cultured banana plantlets as the main factors limiting adoption.

Initial adoption of tissue-cultured banana is highly related to the household's availability of savings, farm equipment, and security of land tenure. Financial capital requirement may prevent poor farmers from adopting a technology if they lack cash to make investments which will not yield a return until after the first harvest (13 months after planting in the case of banana). Farming households that engage in informal employment and trading are more likely to adopt tissue-cultured banana. Engagement in service provision is negatively related to growing tissue-cultured banana, probably because it engages household labour on a full time basis. Consequently, the medium-resource endowment farming households have the largest proportion of households growing tissue-cultured banana. This is most likely because farming households in this category have savings from farm produce sales as they produce both for own consumption and the market. They also earn extra income from non-formal employment. Medium-resource endowment farming households have a high farm equipment value, the main part of which is contributed by water storage tanks from which they irrigate the tissue-cultured banana plantlets. They also have secure land tenure through registration and title deeds which gives them confidence to invest in a perennial crop. Adoption of tissue-cultured banana has implications for the gender division of labour and labour allocations may also have to be renegotiated with the adoption of tissue-cultured banana.

Continued use of tissue-cultured banana is influenced by more factors. The age of the household head and land tenure are important probably because older household heads have more user rights and can easily make decisions on land use. Availability of labour also becomes an important factor as demonstrated by the positively significant effect of household size. Reduction in labour and, probably, the associated loss of income through the death of an adult member reduces the household's likelihood of continuing to use tissue-cultured banana plantlets. This implies that HIV/AIDS-affected households that have lost adult household members cannot afford to invest in tissue-cultured plantlets for further expansion of their banana plots. Engagement in formal employment, which has a negative influence on initial adoption, becomes important for continued use of tissue-cultured plantlets, perhaps because the cash from this activity is used for the purchase of tissue-cultured plantlets and hiring labour.

Continued use of tissue-cultured plantlets relies on contacts with extension agents. Unavailability of the tissue-cultured plantlets is one of the limiting factors to adoption mentioned by farming households. Thus households that have contact with extension agents have access to the planting materials.

Farming households' aspirations in the sample population are to ensure that they meet basic needs such as adequate food, clothing, and shelter, educate the children, maintain good health for family members and increase farm output by purchasing inputs and hiring labour. Both HIV/AIDS-affected and non-affected households farming tissue-cultured banana reported an increase in banana production, income and food which they attributed to the adoption of the technology. However, the proportion of affected households that reported an increase in these benefits was higher. This is probably because these responses were self-reported and therefore subjective. HIV/AIDS-affected households could probably register and appreciate small changes more than non-affected households.

The extra income earned by the sale of increased banana production in both affected and non-affected households is used first to ensure food security and secondly to pay for the education of the children. It is also used to purchase items such as clothing, utensils, and other household items and pay for medical bills, thus contributing to the welfare of the households.

For the technology to benefit resource-poor farming households, who constitute the majority of the population, efforts must be made to increase their access to financial capital or to lower the cost of tissue-cultured plantlets. The laboratories producing tissue-cultured plantlets could reduce the cost of the plantlets by adopting cheaper alternatives to expensive inputs and infrastructure which would reduce the unit cost of plant production. However, the low cost options should lower the cost of production without compromising the quality of the plantlets.

Irrigation is also an essential requirement that influences the adoption of tissue-cultured bananas. Thus there is need to supplement the rain-fed production systems by providing water at the right time and in the right amounts. Poor farming households need to be facilitated to acquire and utilize simple-to-use inexpensive technologies such as the 'bucket and drip kit', treadle pumps and water-harvesting techniques.

8.2 A livelihood approach to agricultural innovation

8.2.1 Agricultural innovation

One of the objectives of this study was to contribute to the livelihood theory by assessing the relevance of agricultural technologies for rural households who are confronted with HIV/AIDS-induced shocks and stress in generating their livelihood. The livelihood approach was used as it has been found by several researchers to provide a relevant framework for identifying the vulnerabilities and susceptibilities of rural households to HIV/AIDS (Loevinsohn & Gillespie, 2003; FAO, 2003; Barnett & Whitesand, 2002; Haddad & Gillepsie, 2001). A few authors have used the approach to study the impact of agricultural innovations on farming households (Adato & Meinzen-Dick, 2002; Kerr & Kolavalli, 1999). The approach provides a method of studying the multiple and interactive influences on livelihoods by incorporating many considerations that are often not included in agricultural technology adoption studies. Adoption studies that are based on the innovation theory over- emphasize the importance of the communication process, thereby overlooking other equally important livelihood factors such as the assets of farming households, characteristics of people, and the endogenous and exogenous environment that influence access to assets and the vulnerability context within which households operate. Understanding the full picture of people's livelihoods can help develop technologies that complement and strengthen the complex livelihood strategies of farming households. The need to explore people's livelihood becomes particularly critical in a situation of HIV/AIDS. The loss of family labour due to a HIV/AIDS-related death in the household means that labour becomes a limiting factor in agricultural production. Hence labour-saving technology may indeed be appropriate for HIV/AIDS-affected. However, it does not necessarily follow that the appropriate policy response for agricultural innovation is to focus exclusively on labour-saving agricultural technology. Rather, agricultural research should also develop affordable technologies that can assure household food security, as well as cash income to pay for school fees and other basic necessities. It is therefore important to not only look at conventional indicators such as land or income, but also at other factors that influence farming households' ability to utilize agricultural technologies. For example, the results of this study indicate that the size of land may not always be an important determining factor and other that factors such as human capital in terms of health, skills and knowledge are more crucial now in the era of HIV/AIDS. We have seen that farming households affected by HIV/AIDS have increasing constraints on their labour and financial capital. As such they may not have the luxury of engaging themselves in long-term development efforts that do not yield a quick return. Additionally, there is a diverse level of asset ownership and livelihood activities among farming households. This diversity must be taken into account by development agents and policy-makers for effective targeting of agricultural technology development.

Farming households in the study area engaged in a wide range of simultaneous livelihood activities and strategies, with agriculture being only one of the several income sources. Even in agricultural production diversity exists among the farming households. The low resource-endowment households are basically practicing subsistence farming and their own labour to earn extra income. Tripp (2001a:485) is of the opinion that such households should be 'helped to find reasonable exits from agriculture, and investment in technology generation on their behalf maybe displaced'. Unfortunately, these households form the majority of the farming population in the study area and they rely on agriculture for household food security. They have low levels of education and little financial capital. This affects their capacity to access high-return non-farm activities. The medium resource-endowment category consists of semi-subsistence farmers. They are engaged in a diversified portfolio of income generating activities but mostly rely on agriculture for household food security. The high resource-endowment category is composed of full-time farming households who produce both for the market and home consumption. According to Tripp, and I concur with his suggestion, these two categories of farming households could be assisted in adopting technologies that enable them to become viable commercial enterprises. These three farming household typologies are comparable to Doward's (2006) three typologies of livelihood strategies that could be used in development. These include:

1. 'Hanging in' strategy whereby people engage in activities to maintain livelihood and ensure that they do not suffer too much in adverse circumstances.
2. 'Stepping up' where people expand their current activities to increase production and income to improve livelihoods.
3. 'Stepping out' where people engage in activities to accumulate assets so that when they have accumulated enough then these assets can provide a base for moving into other activities that have higher or more stable returns (Doward, 2006: 12).

8.2.2 Policies and institutions

Interventions through policies and by institutions change people's options in pursuing livelihood strategies. The policies related to extension services were considered crucial in influencing farming households' decisions in adoption of tissue-cultured banana. The extension policy in Kenya is shifting away from the conventional top-down approach that focuses on the male farmers towards a more participatory demand-driven approach that addresses concerns of all gender and age categories. This is becoming especially necessary in the face of the HIV/AIDS pandemic as the majority of HIV/AIDS-affected households in the rural areas are likely to be middle-aged widows, often caring for young orphaned children. However, the extension system is unable to effectively implement these policies at the farm household level due to budgetary and human capacity constraints. To overcome these constraints the extension systems may require adopting integrated and interactive approaches to deal with the new challenges presented by HIV/AIDS by involving stakeholders with comparative advantage in relevant areas of expertise. In addition, the extension staff may require new skills in handling the changing needs of their clientele. The agricultural extension staff is especially well-placed to provide nutritional information. Focus group discussions with people living with HIV/AIDS (PLWA) revealed that although they know about the importance of proper nutrition, knowledge on the appropriate food crops to grow and how to prepare them in order to preserve the nutrients is lacking. Therefore there is the need to provide technical information that not only increases quantity but also diversity and quality of food.

The limited access to formal credit compels the households to rely on informal credit sources which are most of the time expensive and provide amounts that are often too small for substantial investment. This means that poor households are cannot afford the inputs (tissue-cultured plantlets) required in technology adoption, which further widens the gap between the poor and the rich. Another institution that may influence adoption of agricultural technology is the market. In the case of bananas in Kenya currently, marketing is not yet a constraining factor and an increase in production in the country can easily be absorbed in the local markets (see also Qaim, 1999).

8.2.3 Vulnerability context

This study looked at the vulnerability presented by the effects of HIV/AIDS on farming households' livelihoods and the agricultural extension services. The adoption of tissue-cultured banana has the potential of improving livelihood outcomes and reducing the vulnerability of farming households. This is mainly through improved productivity which leads to increased income, food security and general household welfare. However, adoption is also associated with financial capital and labour costs that HIV/AIDS-affected households may not be able to bear. This emphasizes the need for all stakeholders (policy-makers in agriculture, researchers, extension agents, development agents and farmers) to adopt the 'HIV/AIDS lens' concept (Gillepsie & Kadiyala, 2005; Loevinsohn & Gillespie, 2003) in examining how any new agricultural intervention may reduce or increase the risks of farming households. The HIV/AIDS lens is a conceptual tool intended to reveal the potential contribution (negative or positive) of an intervention to HIV/AIDS prevention and mitigation. The use of the lens will answer questions on how the policy, program or project might increase household's susceptibility to HIV infection and how it might increase their vulnerability or resilience. To some extent, this study applied the lens in analysing the situations at the farm household level. However, a more rigorous application of the concept from the micro- to macro-level will result in policies, program and projects prioritized according to their potential contribution to HIV/AIDS-affected farm household's livelihood.

8.2.4 Gender perspectives on livelihood and HIV/AIDS impacts

Female-headed households were found to be disadvantaged in several ways. For one, the majority of the female-headed households are found in the low-resource endowment category and slightly over half of them are HIV/AIDS-affected. These households have a significantly higher dependency ratio and a higher incidence of labour shortage despite the fact that their size is larger than that of non-affected households. This is because the extra members joining the household after the death of an adult are usually orphaned children who are dependants. Secondly, female household heads have lower level of education than male household heads. This implies that their chances of getting formal employment are lower. Some of the other household members have therefore opted to migrate to urban areas in search of employment. Thirdly, they have little financial capital as they have no savings and their access to informal credit provided by community organisation is limited as they also have few household members engaged in these groups. In addition, female-headed households have significantly lower physical capital in terms of household assets which they could sell to get cash in times of crisis.

Female-headed households have few options of earning extra income. The majority are selling out their own labour on a daily basis. Others have resorted to selling cows which leads to loss of an important source of livelihood for the family. The option of renting-out land taken by other HIV/AIDS-affected households is risky for female-headed households. This is because widows may be reluctant to rent-out land for fear of losing rights, if they have such rights. Migration, an important livelihood strategy undertaken in female-headed HIV/AIDS-affected households, renders households more vulnerable to HIV/AIDS infection as the migrants are usually exposed to conditions that result in HIV infection. They eventually return to the rural homes as carriers of the infection, thus spreading the disease in the rural areas.

Although banana is traditionally considered a woman's crop, the adoption of tissue-cultured banana with the accompanying prospects of increase in yield and value could change this. This finding is not unique to this study as there are various examples where the value of a crop considered being female increases; men take advantage of it instead of women benefiting from this increase (Doss, 2001; Grey & Kevane, 1999). Grey and Kevane quoting an example by Coquery-Vidrovitch (1994) reports a case in Benin where men abandoned the production of yams in favour of oil palm and yams became a woman crop. Later the women recognized the superiority of cassava over yams but later men again asserted control with the availability of new processing techniques for cassava. At the time of study access to and control over banana income still seemed to be in the hands of women, but men were showing keen interest in the crop as it increasingly becomes a main income earner. We also noted initial reluctance by women to grow tissue-cultured bananas in households that had small pieces of land due to competition with food crops. Women in these households were more concerned with achieving food security than income generation. This finding concurs with Niehof's observation that 'women are key actors in achieving food security for their household because it is part of their reproductive role' (Niehof, 2003: 65). Women were also resisting tissue-cultured banana due to the added labour burden of watering. These issues pose a problem for attempts to raise the welfare of women through technical change in 'their' crops. This is because, as Niehof argues, 'because of their reproductive responsibilities and the tasks in the food system that ensues from these, women evaluate resources and assets differently from men. They will have their own classifications and socio-cultural ordering of their environment, in which the underlying value is food security' (Niehof, 2003: 65).

8.3 Policy implications

The results of this study raise important considerations for policy makers and suggest exploration in several areas. The diversity of farming households in terms of assets and activities should play a major role in shaping policy on agricultural technology development. Policy makers should seek ways that will enable poor-resource farming households to move out of agriculture while ensuring household food security. This could be done by enabling medium- and high-resource farming households adopt technologies that will increase production and create jobs locally through hiring labour.

The credit provided by informal networks is helpful in taking the households over difficult times. It also enables them to invest and in the process build up savings that can help lift them out of poverty. In the informal sector there seems to be scope for microfinance institutions that intermediate between local savers and borrowers.

A land policy that is clear on rules governing land rental is required. Developing such a policy will assist HIV/AIDS-affected households earn revenues from renting land that would otherwise go unutilized. This will enable them to raise cash for various household needs including purchasing of farm inputs required in technology adoption.

Technology development should pay attention to the crops that women grow, to avoid social relations and institutions by which gender inequality is perpetuated over time. This may involve ensuring that women continue to have access and control over their labour outputs on these crops as well as ensuring household food security. Besides, access to water and energy sources must be improved, particularly given that access relates to activities that are socially determined to be the responsibility of women. This responsibility has a direct effect on women's time for productive as well as reproductive roles. Furthermore, water is an essential component in the adoption of tissue-cultured banana and most improved crop varieties and water management policies that ensure access to water both for domestic consumption and irrigation need to be pursued. Consequently, development practitioners will have to take into account women's opinion and constraints in introducing interventions and solutions to agricultural problems.

The burden of coping with the impact of HIV/AIDS is shifting on to communities. Several voluntary community initiatives to support AIDS orphans and the elderly have been instigated in the study area mainly providing food and clothing. These community organizations provide the government and the civil society the opportunity to build on these initiatives to help make them more effective and sustainable in mitigating HIV/AIDS impact.

The policies on the agricultural extension system indicate that the government recognizes the crucial role that the institution plays in agricultural development. However, given the prevailing human capacity and budgetary constraints, a new approach needs to find ways and means of doing more with less. This will involve identifying the critical new and expanded needs of farming households, applying interactive training methods to produce effective results. New strategic approaches will be needed to meet the extension needs of AIDS-affected families, targeting especially the resource-poor, female-headed, and grand-parent households. In this extension service system new farmer education and training activities will need to include AIDS education, counselling, and nutritional education in addition to the conventional information on farm improvement. The main targets in this system will be meeting the specific needs of the more vulnerable AIDS-affected and the poorest of households to ensure their production does not collapse and make them less vulnerable to food insecurity.

With extending public education about HIV/AIDS, it is expected that AIDS-related stigmatization will be reduced. However, for effective HIV/AIDS management initial testing for the HIV/AIDS is essential. If people do not know their status they cannot effectively manage the disease. Stigmatization is associated with delays in HIV/AIDS testing by people who are at high risk for infection. During these delays when HIV positive people do not know of their sero-positive status, people may be engaging in risky behaviour and thus unknowingly transmitting HIV to others. It is therefore imperative to continue with public education and promotion of voluntary testing. In addition, voluntary testing needs to be accompanied with affordable ARV drugs which will prolong people's life and also reduce stigmatization.

Finally, we recommend that future studies should adopt a longitudinal methodological approach which will enable researchers to capture the long-term impacts of HIV/AIDS as well as address the various dimensions and dynamics of technology adoption. It would be particularly important to return to the study area after another two to three years, to further explore the impacts of HIV/AIDS and tissue-cultured banana technology on the livelihoods of the population. To follow-up on this I intend to link up with development partners that are still promoting the tissue-cultured banana and other agricultural technologies in the region.

References

Adato, M. & R. Meinzen-Dick (2002). Assessing the impact of agricultural research on poverty using the sustainable livelihoods framework. *FCND Discussion Paper 128* International Food and Policy and Research Institute (IFPRI), Washington D.C. USA.

Adesina, M.M. & M.M. Zinnah (1993). Technology characteristics, farmers' perceptions and adoption decisions: A Tobit model application in Sierra Leone *Agricultural Economics* 9(4): 297-311.

Agarwal, B. (1985). Women and technological change in agriculture: the Asian and African experience. In: Ahmed I. (Ed.) *Technology and Rural Women, Conceptual and Empirical Issues.* London, George Allen & Unwin, pp. 67-114.

Ainsworth, M. & I. Semali (1998). Who is most likely to die of AIDS? Socio-economic correlates of adult deaths in Kagera Region, Tanzania. In: Aisnworth, M., L. Fransen and M. Over (Eds.) *Confronting Aids: Evidence from the Developing World*, Brussels: European Commission, pp. 95-109.

Ali, A. (2005). *Livelihood and Food security in Rural Bangladesh.* Ph.D. Thesis Wageningen University.

Aliber, M. & C. Walker (2006). The impact of HIV/AIDS on land rights: Perspectives from Kenya. *World Development* 34(4): 704-727.

Armstrong, J. (1995). Uganda's Crisis: its implication for development. *Discussion Papers No. 298.* Washington, DC: World Bank.

Baerends, A.E. (1994). *Changing Kinship, Family and Gender Relationships in Sub-Saharan Africa,* Leiden: Women and Autonomy Centre (VENA), Leiden University.

Barnett, T. & A. Whiteside (2002). *AIDS in the Twenty-First Century. Disease and Globalization,* Palgrave: Macmillan.

Baylies, C. (2002). The impacts of AIDS on rural households in Africa: a shock like any other? *Development and Change* 33(4): 611-632.

Beegle, K. (2003). Labour effects of adult mortality in Tanzanian households, *Policy Research Working Paper* Series 3062, Washington D.C.: The World Bank.

Beets, C.W. (1990). *Raising and Sustaining Productivity of Smallholder Farming Systems in the Tropics,* A Handbook of Sustainable Agricultural Development. AgBe Publishing.

Berry, S. (1993). *No Condition is Permanent. The Social Dynamics of Agrarian Change in Sub-Sahara Africa,* The University of Wisconsin Press 114 North Murray.

Berry, S. (2002). Debating the land question in Africa, *Comparative Studies in Society and History,* 44 (4): 638–68.

Biggs, S.D. (1990). A multiple source of innovation model of agricultural research and technology promotion. *World Development* 18: 1481-1499.

Bindlish, V. & R.E. Evenson (1993). *Evaluation of the Performance of T & V Extension in Kenya,* World Bank Technical Paper No. 208, Africa Technical Department Series, Washington D.C.: The World Bank.

Blaikie, P.M., I.D. Cannon & B. Wisner (1994). *At Risk: Natural Hazards, People's Vulnerability and Disasters.* London/New York: Routledge.

Blanc, A. (2000). *The Relationship Between Sexual Behaviour and Level of Education in Developing Countries.* Report prepared for UNAIDS, Geneva: UNAIDS.

Bollinger, L., J. Stover & D. Nalo (1999): 'The Economic Impact of AIDS in Kenya' Washington DC: The Futures Group International in collaboration with: Research Triangle Institute (RTI) and The Center for Development and Population Activities), USAID, http://web.worldbank.org/WBSITE/EXTERNAL/COUNTRIES/AFRICAEXT/EXTAFRHEANUTPOP/EXTAFRREGTOPHIVAIDS/0,,contentMDK:20435845~pagePK:34004173~piPK:34003707~theSitePK:717148,00.html (accessed on 11th January 2007).

Boserup, E. (1970). *Women's Role in Economic Development*. Aberdeen: Aberdeen University Press.

Bouman, F.J.A (1995). Rotating and accumulating savings and credit associations: A development perspective, *World Development* 23(3): 371-84.

Brad D.J. (2002). Migration and agricultural change: The case of smallholder agriculture in highland Ecuador, *Human Ecology* 30(4): 523-550.

Brons, J., T. Dietz, A. Niehof & K. Witsenburg (2007). Dimensions of vulnerability of livelihoods in less-favoured areas: Interplay between the individual and the collective. In: Ruben R., J. Pender and A. Kuyvenhoven (Eds.) *Sustainable Poverty Reduction in Less-Favoured Areas,* Wallingford: CAB International [In press].

Bruce J. & C.B. Lloyd (1995). Finding the ties that bind: beyond headship and household. In: Arno van der Avort, Kees de Hoog and Pieter Kalle (Eds.) *Single Parent Families*. The Hague: The Netherlands Family Council, pp. 60-90.

Bryceson, D.H. (1999). Sub-Saharan Africa betwixt and between: Rural livelihood practices and policies, ASC *Working Paper 43,* Leiden: Africa Studies Centre.

Buvinic, M. & G.R. Gupta (1997). Female-headed households: are they worth targeting to reduce poverty in developing countries? Economic *Development and Cultural Change* 45(2): 259-80.

Byerlee, D. (1996). Modern varieties, productivity and sustainability, *World Development* 24(4): 697-718.

Carney, D. (1998). Sustainable rural livelihoods. What contribution can we make? *Paper presented at the DFID Natural Resources Advisers' Conference,* July 1998. London: DFID.

Carr, M. (1985). Technologies for rural women: impact and dissemination. In: Ahmed I. (Ed.) *Technology and Rural Women, Conceptual and Empirical issues*. London: George, Allen & Unwin.

Carswell, G. (1997). Agricultural intensification and rural sustainable livelihoods: A 'think piece'. *IDS Working Paper 64* University of Sussex: Institute of Development Studies. *http://www.ids.ac.uk/ids/bookshop/wp/Wp64.pdf* (accessed on 10th June, 2006).

Carter, M.R. & C.B. Barrett (2006). The economics of poverty traps and persistent poverty: an asset-based approach. *Journal of Development Studies*, 42(2): 178-199.

Chambers, R. (1983). *Rural Development: Putting the last first*, London/New York: Longman.

Chambers, R. (1990). Vulnerability, coping and policy, *IDS Bulletin* 20 (2): 1-7.

Chambers, R. & G. Conway (1992). Sustainable livelihoods: Practical concepts for the 21st century *IDS Discussion Paper* 296, University of Sussex: Institute of Development Studies.

Chant, S. (1997). *Women-Headed Households: Diversity and Dynamics in the developing World.* Basingstoke: Macmillan.

Chapoto, A. & T.S. Jayne (2005). socio-economic characteristics of individuals afflicted by AIDS-related prime-age mortality in Zambia. *Policy Synthesis No.10. Food Security Research Project – Zambia.* Ministry of Agriculture and Cooperatives, Agricultural Consultative Forum, Michigan State University. *http://www.aec.msu.edu/agecon/fs2/zambia/index.htm* (accessed on 9 September 2006).

CIMMYT Economics Program (1993). *The Adoption of Agricultural Technology: A Guide for Survey Design,* D.F.: CIMMYT.

Clough, S.M. (1998). *Mau Mau Memoirs: History, Memory and Politics.* Boulder, Colorado: Lynne Reinner Publishers.

Cohen, D. (2002). *Human Capital and the HIV Epidemic in sub-Saharan Africa,* Working Paper 2, ILO Programme on HIV/AIDS and the World of Work, Geneva. http://www-ilo-mirror.cornell.edu/public/english/protection/trav/aids/publ/wp2_humancapital.pdf (Accessed on July 12th 2006).

Conley, T.G. & C.R. Udry (2001). Learning about a new technology: pineapple in Ghana. *American Journal of Agricultural Economics* 83(3): 668-673.

De Bruyn M. (1992). Women and AIDS in developing countries. *Social Science & Medicine* 34(3): 249-62.

De Haan, A. (1999). Livelihoods and poverty: the role of migration. A critical review of the migration literature, The *Journal of Development Studies* 36(2): 1-47.

De Janvry, A. & E. Sadoulet (2002). World poverty and the role of agricultural technology: Direct and indirect effects. *The Journal of Economic Development Studies* 38(4): 1-26.

Dewees, P.A. (1995). *Social and Economic Incentives for Smallholder Tree Growing A Case Study from Murang'a District, Kenya.* Community Forestry Case Study Series, Rome: FAO.

DFID (2001). Sustainable Livelihoods Guidance Sheets, Department for International Development, http://www.livelihoods.org/. (Accessed on 12th June 2005).

Doss, C.R. (2001). Designing agricultural technology for African women farmers: Lessons from 25 years of experience. *World Development* 29(12): 2075-2092.

Doss, C.R. (2002). Men's crops? Women's crops? The gender patterns of cropping in Ghana. *World Development* 30(11): 1987-2000.

Doss, C.R. (2006). Analyzing technology adoption using microstudies: Limitations, challenges, and opportunities for improvement *Agricultural Economics* 34: 207-219.

Doss C.R. & L.M. Morris (2001). How does gender affect the adoption of agricultural innovations? The case of improved maize technology in Ghana. *Agricultural Economics* 25(1): 27-39.

Doss, C., W. Mwangi, H. Verkuijl & H. De Groote (2002). *Adoption of maize and wheat technologies in East Africa: Synthesis of East African adoption studies.* CIMMYT Economics Working Paper 02-04. Mexico, D.F.: CIMMYT.

Doward, A (2006). Poverty reduction, progress and processes - some personal reflections. Focus on poverty reduction, *WYE connect*, pp. 11-14.

Doward, A., J. Kydd, J. Morrison & I. Urey (2004). A policy agenda for pro-poor agricultural growth. *World Development* 32(1): 73-89.

References

Drimie, S. (2002). The Impact of HIV/AIDS on Land: Case Studies from Kenya, Lesotho and South Africa. *A Synthesis Report Prepared for Southern African Region Office of the Food and Agricultural Organisation of the United Nations* (FAO) Rome: FAO.

Du Guerny, J. (2002). Agriculture and HIV/AIDS. *UNDP SEAHIV/FAO/EASE* International. *http://www.hiv-development.org/publications/Agriculture.htm* (Accessed on 13th May 2005.).

Ekaas, S. (2003). *Gender HIV/AIDS and Food Security.* Paper Prepared for the ECOSOC Ministerial Roundtable on Gender and Rural Development, New York: UNIFEM.

Ellis, F. (1993). *Peasant Economics: Farm Households and Agrarian Development.* Cambridge: Cambridge University Press.

Ellis, F. (1998). Household strategies and rural livelihood diversification, *Journal of Development Studies* 35(1): 1-38.

Ellis, F. (2000). *Rural Livelihoods and Diversity in Developing Countries,* Oxford: Oxford University Press.

Evans, A. (1989). Gender Issues in Rural Household Economics. *IDS Discussion Paper no. 254,* Brighton: Institute of Development Studies.

Evenson R.E. & G. Mwabu (2001). The effect of agricultural extension on farm yields in Kenya, *African Development Review* 13(1): 1-23.

FAO (1995). *The effects of HIV/AIDS on farming systems in Eastern Africa.*, Rome: FAO.

FAO (1998). Socio-economic impact of HIV/AIDS on agriculture, *Briefing Paper,* Rome: FAO.

FAO (2001). Biotechnology and banana production, intergovernmental group on bananas and tropical fruits, *http://www.fao.org/DOCREP/MEETING/004/Y1896E.HTM* (Accessed on 20th October 2006).

FAO (2003). HIV/*AIDS Food Security and Rural Livelihoods.* FAO, Website.

FAO(2004). FAO Statistical Yearbook 2004. *http://www.fao.org/statistics/yearbook/vol_1_2/pdf/Kenya.pdf* (Accessed on 20th September 2006).

FAO (2006). Gender, Food and Security. *Facts and Figures. http://www.fao.org/GEnder/en/agrib4-e.htm* (Accessed on 20th September 2006).

Feder, G., E. Richard & D. Zilberman (1985). Adoption of agricultural innovations in developing countries: A Survey, *Economic Development and Change* 33(2): 255-298.

Fernandez-Cornejo, J. & W.D. McBride (2002). Adoption of bioengineered crops. *Agricultural Economic Report No. 810,* Washington DC: United States Development of Agriculture Economic Research Service.

Field, A. (2000). *Discovering Statistics Using SPSS for Windows. Advanced Techniques for the Beginner,* London/ New Delhi: Sage Publications.

Forest Action Network (2002). *The Impacts of HIV/AIDS on the Land Issue in Kenya,* Rome: FAO.

Fox, M.P., S. Rosen, B.W. MacLeod, M. Wasunna, M. Bii, G. Foglia & J.L. Simon (2004). The impact of HIV/AIDS on Labour Productivity in Kenya. *Tropical Medicine and International Health* 9(3): 318–324.

Francis, E. (2002). Gender, migration and multiple livelihoods: Cases from Eastern and Southern Africa, *The Journal of Development Studies* 38(5): 167-190.

Gillespie, S. (1989). Potential impact of AIDS on farming systems: A case study from Rwanda. *Land Use Policy* 6(4): 301-312.

Gillespie, S. & S. Kadiyala (2005). HIV/AIDS and food and nutrition security: Interactions and response, *American Journal of Agricultural Economics* 87(5): 1282–1288.

Gillespie S., L. Haddad & R. Jackson (2001). HIV/AIDS food and nutrition security: Impact and actions. *Nutrition Policy paper No. 5* Geneva: United Nations.

Gold, C.S. (2001). Biology and integrated pest management of banana weevil, *Cosmopolites sordidus* (Germar) *Advancing Banana and Plantain R & D in Asia and the Pacific* 10(28): 33 Los Baños: INIBAP-ASPNET.

Gold, C.S. & S. Messiaen (2000). The banana weevil *Cosmopolites Sordidus. Musa Pest Fact sheet* No.4. Montpellier: INIBAP.

Gonsalves, J., T. Becker, A. Braun, D. Campilan, H. De Chavez, E. Fajber, M. Kapiriri, J. Rivaca-Caminade & R. Vernooy (2005). *Participatory Research and Development for Sustainable Agriculture and Natural Resource Management: A Sourcebook.* Volume 1: Understanding Participatory Research and Development. International Potato Centre-Users' Perspectives with Agricultural Research and Development, Laguna, Phillipines and International Development Research Centre Ottawa, Canada.

Gray, L. & M. Kevane (1999). Diminished access, diverted exclusion: women and land tenure in Sub-Saharan Africa. *African Studies Review* 42(2): 15-39.

Greene, J.C., V.J. Caracelli & W.D. Graham (1989). Toward a conceptual framework for mixed-method evaluation designs, *Educational Evaluation and Policy Analysis* 11(3): 255-274.

Guenter, H. (1999). *The Relevance of HIV/AIDS to GTZ-Supported Projects in Kenya,* Workshop Report, GTZ-Kenya, Nairobi, March 1998.

Haddad, L. & S. Gillespie (2001). Effective food and nutrition policy responses to HIV/AIDS: What we know and what we need to know. *Journal of International Development* 13: 487-511.

Hair, J.F. Jr., E.A. Rolf & L.T. Ronald (1987). *Multivariate Data Analysis.* New York: Collier Macmillan Publishers. [Second Edition].

Hardon-Baars, A. (1997). Users' perspectives: Literature review on the development of a concept. *UPWARD Working Paper Series* No. 4, Los Baños, Laguna: UPWARD.

Herek, G.M. (1999). AIDS, stigma and sexual prejudice. *American Behavioural Scientist* 42(7): 1130-1147.

Imai, K., 2003. Is livestock important for risk behaviour and activity choice of rural households? Evidence from Kenya, *Journal of African Economies* 12(2): 271-295.

International Network for Improvement of Banana and Plantain (INIBAP) (2000). INIBAP Annual report 2000, Montpellier: INIBAP.

Jaetzold, R. & H. Schmidt (1983). *Farm management handbook of Kenya.* 4 vols. Nairobi: Kenya Ministry of Agriculture.

Jayne, T.S., M. Villarreal, P. Pingali & G. Hemrich (2004). HIV/AIDS and the Agricultural Sector: Implications for Policy in Eastern and Southern Africa. *Journal of Agricultural and Development Economics* 2(2): 158-181.

Kabeer, N. (1991). Gender, production and well-being: Rethinking the household economy, *IDS Discussion Paper* No 288, Brighton: Institute for Development Studies.

Karanja, A.M. (2002). *Liberalisation and Smallholder Agricultural Development: A Case Study of Coffee Farms in Kenya.* Ph.D. Thesis, Wageningen University.

Kenya Agricultural Research Centre (1999). *Centre Annual Report* 1999 National Horticultural Research Centre Thika, Kenya (Unpublished).

Kerr, J. & S. Kolavalli, (1999). Impact of Agricultural Research on Poverty Alleviation: Conceptual Framework with Illustrations from the Literature *EPTD Discussion Paper* No.56 Washington DC, USA: International Food Policy Research Institute (IFPRI).

Kiriti, T.W., C. Tsidell & C.K. Roy (2003). Female participation in decision making in agricultural households in Kenya: Empirical findings. *International Journal for Agricultural Resources, Governance and Ecology* 2(2): 103-124.

Koopman, J. (1991). Neo-classical household models and modes of household production: Problems in the analysis of African agricultural households. *Review of Radical Political Economics* 23(3/4): 148-173.

Kung'u, J.N. (1998). *Ecology, Distribution and Population Structure of Fusarium oxysporum f.sp. cubenses in Kenya,* PhD Thesis, University of Kent, Canterbury, UK.

Leeuwis, C. & A. van den Ban (2004). *Communication for Rural Innovation: Rethinking Agricultural Extension,* Oxford: Blackwell Science. [3rd Edition].

Lipton, M. (2001). Reviving global poverty reduction: What role for genetically modified plants. *Journal of International Development* 13(7): 823-846.

Loevinsohn, M. & S. Gillespie (2003). HIV/AIDS, Food Security and Rural Livelihoods: Understanding and Responding. *Renewal Working Paper No.2* IFPRI/ISNAR: Lynne Rienner Publishers.

Mahiri, I. & C. Howorth (2001). Twenty years of resolving the irresolvable: approaches to the fuelwood problem in Kenya, *Land Degradation and Development* 12: 205-215.

Maredia, M.K., D. Byerlee & P. Pee (2000). Impacts of food crop improvement research: Evidence from Sub-Saharan Africa, *Food policy* 25: 531-559.

Mbaga-Semgalewa, Z. & H. Folmer (2000). Household adoption behaviour of soil conservation: The case of North Pare and West Usambara Mountains of Tanzania. *Land Use Policy* 17: 321-336.

Mencher, J.P. & A. Okongwu (1993). Where *Did All the Men Go? Women in Cross-Cultural Perspectives,* Boulder: Westview Press.

Menon, R., M.J. Wawer, J.K. Konde-Lule. N.K. Sewanlambo & C. Li (1998). The economic impact of adult mortality on households in Rakai District, Uganda. In: M. Ainsworth, L. Fransen and M. Over (Eds.) *Confronting AIDS: evidence from the development World.* Brussels: European Commission, pp. 325-41.

Miceli, T.J., C.F. Sirmans & J. Kieyah (2001). The demand for land title registration: Theory with evidence from Kenya. *American Law and Economics Review* 3(2): 275-287.

Ministry of Agriculture (MOA) (2004). *District Farm Management Guidelines,* Maragua District, District Agricultural Officer, Makuyu, Kenya (Unpublished).

Moll, H.A.J. (2005). Costs and benefits of livestock systems and the role of non-market relationships, *Agricultural Economics* 32(2): 181-193.

Moser, C.O. (1989). Gender planning in the Third World: Meeting practical and strategic gender needs. *World Development* 17(11): 1799-1825.

Moser, C.O. (1998). The asset vulnerability framework: reassessing urban poverty reduction strategies. *World Development* 26(1): 1-19.

Mtshali, S.M. (2002). *Household Livelihood Security in Rural KwaZulu-Natal, South Africa*. Ph.D. Thesis. Wageningen University.

Müller, T.R. (2004). *HIV/AIDS and Agriculture in Sub-Saharan Africa,* Wageningen: Wageningen Academic Publishers. [AWLAE Series No.1].

Müller, T.R. (2005a). HIV/*AIDS, Gender and Rural Livelihoods in sub-Saharan Africa,* Wageningen: Wageningen Academic Publishers. [AWLAE Series No. 2].

Müller, T.R. (2005b). *HIV/AIDS and Human Development in Sub-Saharan Africa,* Wageningen: Wageningen Academic Publishers. [AWLAE Series No. 3].

Mutangadura, G., D. Mukurazita & H. Jackson (1999). *A Review of Household and Community Responses to HIV/AIDS Epidemic in Rural Areas of sub-Sahara Africa,* UNAIDS Best Practice Paper UNAIDS web site at: www.unaids.org/publications (Accessed on 12th September 2003.).

Muyanga, M. & T.S. Jayne (2006). *Agricultural extension in Kenya: Practice and Policy Lessons,* Tegemeo Working Paper 21. Kenya: Tegemeo Institute of Agricultural Policy Development, Egerton University.

Narayan, D. (1997). *Voices of the Poor: Poverty and Social Capital in Tanzania*. ESSD Studies and Monographs Series, Vol. 20. Washington, DC: The World Bank.

Narayan, D. & D. Nyamwaya (1996). Learning from the poor: A participatory poverty assessment in Kenya.' Environment Department Papers, Participation Series 34. Washington, D.C.: World Bank, Social Policy and Resettlement Division, [Processed].

Narayan, D. & L. Pritchett (1999). Cents and sociability: Household income and social capital in rural Tanzania. *Economic Development and Cultural Change* 47(4): 871-897.

National AIDS Control Council (NACC) (2003). Antenatal Clinical Surveillance Data, Maragua District Hospital. [Unpublished].

Negash, A. & A. Niehof (2004). The significance of Ensete culture and biodiversity for rural household food and livelihood security in South Western Ethiopia, *Agriculture and Human Values* 21: 61-71.

Negatu, W. & A. Prikh (1999). The impact of perception and other factors on the adoption of agricultural technology in the Moret and Jiru Woreda (District) of Ethiopia. *Agricultural Economics* 21: 205-216.

Nguthi, F.N. (1999). Banana production and research in Kenya. *Horticultural News, the Official Journal of HCDA* 18: 7-8.

Niehof, A. (2004). The significance of diversification for rural livelihood systems. *Food Policy* 29: 321-38.

Niehof, A. (2003). Gendered dynamics of food security. In: K. Ulmer (Ed.) *No Security without Food Security - No Food Security without Gender Equality*, Brussels: APRODEV, pp. 60-68.

Niehof, A. & L. Price (2001). *Rural Livelihood, Systems: A Conceptual Framework* UPWARD Working Paper No. 5 Series WU-UPWARD: Wageningen.

Ocholla-Ayayo, A.B.C. (1997). HIV/AIDS Risk Factors and Changing Sexual Practices in Kenya. In: Thomas S. Weisner, Candice Bradley & Philip 1. Kilbride in collaboration with A.B.C. Ocholla-Ayayo, Joshua Akong'a & Simiyu Wandibba (Eds.) *African families and the Crisis of Social Change,* Westport [etc.]: Bergin & Garvey, pp. 110-124.

Onyango, M., F. Nguthi, J. Mutisya & F. Muniu (2004). *Characterization of Banana Cultivars, Production Practices and Constraints of Production for Farmers in Banana Growing areas of Kenya,* Paper presented at the International Congress on Musa: Harnessing Research to Improve Livelihoods – 6-9 July 2004, Penang, Malaysia.

Pennartz, P. & A. Niehof (1999). The *Domestic Domain. Chances, Choices and Strategies of Family Households,* England: Ashgate Publishing Limited.

Place, F. & B. Swallow (2000). *Assessing the Relationships between Propoerty Rights and Technology Adoption in Smallhoder Agriculture: A Review of Issues and Empirical Methods.* CAPRi Working Paper No.2.

Portes, A. (1998). Social capital: Its origins and applications in modern sociology, *Annual Review of Sociology* 24: 1-24.

Putnam, R. (1993). *Making Democracy Work: Civic Traditions in Modern Italy.* Princeton, NJ: Princeton University Press.

Qaim, M. (1999). Assessing the Impact of Banana Biotechnology in Kenya. *ISAAA Briefs* No. 10-1999. Bonn: Published by ISAAA in Collaboration with Zentrum für Entwicklungsforschung/Centre for Development Research, Universität Bonn.

Rankin, K.N. (2002). Social capital, microfinance, and the politics of development, *Feminist Economics* 8(1): 1-24.

Reardon, T., E. Crawford & V. Kelly (1995). *Promoting farm investment for sustainable intensification of agriculture, international Development Paper 18,* East, Lansing: Michigan State University/ Department of Agricultural Economics.

Reardon, T. (1997). Using evidence of household income diversification to inform study of the rural nonfarm labor market in Africa, *World Development* 25(5): 735-747.

Reddy, S.K.V., L. Ngode, J.W. Ssenyonga, W. Wabule, M. Onyango & T.O. Ngoze (1998) Management of pests and diseases of banana in Kenya: A status report. In: E.A. Frison, C.S. Gold, E.B. Karamura & R.A. Sikora (Eds.) *Mobilizing IPM for Sustainable Banana Production in Africa.* Proceedings of workshop on banana IPM held in Nelspruit, 1998 November. Montpellier: INIBAP, pp. 215-223.

Republic of Kenya (1997). *Sessional Paper* No. 4 of 1997 on AIDS in Kenya, Nairobi: Government Kenya.

Republic of Kenya (1999). *National Poverty Eradication Plan 1999-2015,* Nairobi: Department of Development Co-ordination, Ministry of Planning and Development.

Republic of Kenya (2001a). *Poverty reduction Strategy Paper 2001-2004,* Nairobi: Ministry of Finance.

Republic of Kenya (2001b). *National Agricultural Extension Policy,* Nairobi: Ministry of Agriculture and Rural Development.

Republic of Kenya (2003a). *Economic Recovery Strategy for Wealth and Employment Creation 2003-2007.* Ministry of National Planning and Development, Nairobi: Kenya.

Republic of Kenya (2003b). *Demographic and Health Survey 2003,* Nairobi: Kenya.

Republic of Kenya (2003c). *Millennium Development Goals, Progress report for Kenya 2003,* Nairobi: Government of Kenya/UN.

Republic of Kenya (2004a). *National Agricultural Extension Policy,* Work Plan July 2004-2005, Nairobi: Ministry of Agriculture and Rural Development.

Republic of Kenya (2004b). *Strategy for Revitalising Agriculture 2004-2014*, Nairobi: Ministry of Agriculture and Ministry of Livestock and Fisheries Development.

Republic of Kenya (2005a). *Review of the National Agricultural Extension Policy (NAEP) and its Implementation.* Volume II Main Report and Annexes, Ministry of Agriculture and Ministry of Livestock and Fisheries Development, Nairobi April 2005.

Republic of Kenya (2005b). *Aids in Kenya, Trends, Intervention and Impact.* Nairobi: Ministry of Health [7th Edition].

Ringheim, K. (1995). Ethical issues in social science research with special reference to sexual behaviour research. *Social Science & Medicine* 40(12): 1691-1697.

Ritchie, J. (2003). The applications of qualitative methods to social research. In: Ritchie, J. and Lewis, J. (Eds.) *Qualitative Research Practice, A Guide for Social Science Students and Researchers.* London: Sage Publications Ltd., pp. 24-46.

Robinson, J.C. (1996). *Bananas and Plantains*, Wallingford: CAB International, UK.

Rogers, E.M. (1995). *The Diffusion of Innovations*, New York: Free Press [5th Edition].

Rudie, I. (1995). The significance of 'eating': co-operation, support, and reputation in Kelantan Malay households. In: Wazir Jahan Karim (Ed.) *'Male' and 'Female' in Developing Southeast Asia,* Oxford: Berg Publishers, pp. 227-247.

Rugalema, G. (1998). *HIV/AIDS and the Commercial Agriculture Sector of Kenya: Impact, Vulnerability, Susceptibility and Coping Strategies.* Rome: FAO/SDRE (Extension, Education and Communication Service), Research, Extension and Training Division.

Rugalema, G. (1999). *Adult mortality as entitlement failure, AIDS and the related crisis of rural livelihood in a Tanzanian village* The Hague: Institute of Social Studies.

Rukazambuga, N.D.T.M., C.S. Gold & S.R. Gowen (1998). Yield loss in East African highland banana (*Musa* spp., AAA-EA group) caused by the banana weevil, *Cosmopolites sordidus* Germar, *Crop Protection* 17: 581-589.

Russell, B.H. (2002). *Research Methods in Anthropology. Qualitative and Quantitative Methods*, New York: Oxford. [Third Edition].

Salasya, B.D.S. (2005). *Crop Production and Soli Nutrient Management: An Economic Analysis of Households in Western and Central Kenya.* Ph.D. Thesis, Wageningen University.

Scoones, I. (1998). Sustainable Rural Livelihoods. A Framework for Analysis, *IDS Working Paper No. 72*, Brighton: IDS.

Scoones, I. & J. Thomson (1994). Knowledge, power and agriculture-towards a theoretical understanding, In: Scoones and Thompson, J. (Eds.) *Beyond Farmer First. Rural Peoples' Knowledge and Extension Practice,* London: Intermediate Technology Publications, pp. 16-32.

Scrimshaw, C.M.S. (1990). Combining quantitative and qualitative methods in the study of intrahousehold resource allocation, In: Rogers and Schlossman, (Eds.) *Intra-household Resource Allocation: Issues and Methods for Development, Policy and Planning. Papers Prepared for the Workshop on Methods of Measuring Intrahousehold Resource Allocation,* Gloucester, Massachusetts: United Nations University Press, pp. 86-88.

Sen, A.K. (1990). Gender and co-operative conflicts, In: Tinker, I. (Ed.) *Persistent Inequalities,* New York: Oxford University Press, pp. 123-149.

Smith, J., A.D. Barau, A. Goldman & J.H. Mareck (1994). The role of technology in agricultural intensification: The evolution of maize production in the Northern Guinea Savannah of Nigeria. *Economic Development and Cultural Change* 42(3): 537-554.

Smucker, T.A. (2002). Land tenure reform and changes in land use and land management in semi-arid Tharaka Kenya. *Land Use Change Impacts and Dynamics (LUCID) Project Working Paper 11* Nairobi: International Livestock Research Institute.

Society for International Development (SID) (2004). *Pulling Apart: Facts and Figures on Inequality in Kenya.* Nairobi: Kenya.

Spring, A. (2000). Commercialisation and women farmers in Kenya. In: Anita Spring (Ed.) *Women Farmers and Commercial Ventures. Increasing Food Security in Developing Countries,* Boulder: Lynne Rienner publishers, pp. 317-341.

Stokes, C.S. (2003). Measuring Impacts of HIV/AIDS on Rural Livelihoods and Food Security, *http://www.fao.org/sd/2003/PE0102a*_en.htm (Accessed on 24th July 2006.).

Stover, R.H. & N.W. Simmonds, (1987). Bananas. Longman, London.

Swift, J. 1989. Why are rural people vulnerable to famine? *IDS Bulletin* 20(2): 8-15.

Thirtle, C., L. Lin & J. Piesse (2003). The impact of research-led agricultural productivity growth on poverty reduction in Africa, Asia And Latin America, *World Development* 31(12): 1959-1976.

Tibaijuka, A.K. (1997). AIDS and economic welfare in peasant agriculture: Case Studies from Kagabiro village Kagera Region, Tanzania, *World Development* 25(6): 963-975.

Tiffen, M., M. Mortimore & F. Gichuki (1994). *More People, Less Erosion: Environmental Recovery in Kenya.* Chichester: Wiley.

Tittonell, P., B. Vanlauwe, P.A. Leffelaar, E.C. Rome & K.E. Giller (2005). Exploring diversity in soil fertility management of smallholder farms in western Kenya. I. Heterogeneity at region and farm scale, *Agriculture, Ecosystems and Environment* 110: 149-165.

Toner, A. (2003) Exploring sustainable livelihoods approaches in relation to two interventions in Tanzania, *Journal of International Development* 15: 771–781.

Topouzis D. & J. de Guerny (1999). Sustainable agricultural/rural development and vulnerability to the AIDS epidemic, UNAIDS/FAO Best Practice Paper, December 1999. *www.unaids.org* (accessed in November, 2004).

Topouzis, D. (2000). Measuring *the Impact of HIV/AIDS on the Agricultural Sector in Africa.* Joint and Co-Sponsored Program on AIDS, Geneva: UNAIDS.

Tripp, R. (2001a). Agricultural technology policies for rural development. *Development Policy Review* 19(4): 479-489.

Tripp, R. (2001b). Twixt cup and lip-biotechnology and resource-poor farmers. *Nature Biotechnology* Vol. 19 February, *http://biotech.nature.com* (accessed on 12th September 2006).

UNAIDS (1999). *Gender and HIV/AIDS: Taking Stock of Research and Programmes,* UNAIDS: Geneva.

UNAIDS (2000). *Report on the Global HIV/AIDS Epidemic,* Geneva: UNAIDS.

UNAIDS (2002). *Report on the Global HIV/AIDS Epidemic,* Geneva: UNAIDS.

UNAIDS (2003). *Report on the Global HIV/AIDS Epidemic,* Geneva: UNAIDS.

UNAIDS (2004). *Epidemiological Fact Sheets on HIV/AIDS and Sexually Transmitted Infections.* Geneva: UNAIDS/UNICEF/WHO.

UNAIDS (2006). *Report on the global AIDS epidemic.* Geneva: UNAIDS.

UNFPA (2005). *State of World Population 2005, The Promise of Equality, Gender Equity, Reproductive health and the Millennium Development Goals,* New York: UNFPA.

Urassa, M., T.J. Boerma, R. Isingo, J. Ngalula, J. Ng'weshemi, G. Mwaluko & B. Zaba (2001). The impact of HIV/AIDS on mortality and household mobility in rural Tanzania, *AIDS* 15(5): 2017-2023.

Ventura-Dias, V. (1985). Modernisation, production organisation and rural women in Kenya, In: Ahmed I (Ed) *Technology and Rural Women, Conceptual and Empirical Issues.* London George Allen & Unwin, pp. 157-210.

Wambua, S. (2004). Water Privatization in Kenya, Global Issue Papers, No. 8, Heinrich Böll Foundation. *http://www.boell.de/downloads/global/Water%20Privatisation%20in%20Kenya.pdf* (accessed on 24th June, 2006).

Wambugu, F.M. (2001). *Modifying Africa. How biotechnology can benefit the poor and hungry, a case study from Kenya.* Nairobi: Kenya.

Wambugu, F.W. & R.M. Kiome (2001). *The Benefits of Biotechnology for Small-Scale Banana Producers in Kenya.* Ithaca, New York: ISAAA.

Watts, M.J. & H.G. Bohle (1993). Hunger, famine and space of vulnerability, *GeoJournal* 30(2): 117-125.

White, J. & E. Robinson (2000). HIV/*AIDS and Rural Livelihoods in Sub-Saharan Africa,* Chatham: Natural Resources Institute [Policy Series 6].

Whitehead, A. (1985). Effects of technological change on rural women: A review of analysis and concepts. In: Ahmed, I. (Ed) *Technology and Rural Women, Conceptual and Empirical Issues.* London: George Allen & Unwin, pp. 27-64.

Whitehead, A. & D. Tsikata (2003). Policy discourses on women's land rights in sub-saharan africa: The implications of the re-turn to the customary. *Journal of Agrarian Change* 3(1/2): 67-112.

Wiegers, E., J. Curry, A. Garbero & J. Hourihan (2006). Patterns of vulnerability to AIDS impacts in Zambian Households *Development and Change* 37(5): 1073-1092.

Woolcock, M. & D. Narayan (2000). Social capital: Implication for development theory, research and policy, *The World Bank Research Observer* 15(2): 225-49.

World Bank (1997). *Confronting AIDS: Public Priorities in a Global Epidemic.* New York: Oxford University Press.

World Bank (1999). *Confronting AIDS: Public Priorities in a Global Epidemic.* Revised edition, New York: Oxford University Press.

World Bank (2001). Social Capital for Development, Online August 7(3 September 2001).

Yamano, T. & T.S. Jayne (2004). Measuring the impacts of working-age adult mortality on small-scale farm households in Kenya, *World Development* 32(1): 91-119.

Annex

Table AI.I. Description and summary statistics of the variables used in the logistic regression analysis of farming households' adoption of tissue-cultured banana in Maragua.

Variable	Adopters (N=67)		Non-adopters (N=187)	
	Mean	SD	Mean	SD
Human capital				
Household size (persons)	5.8	2.7	5.7	2.5
Age of household head (years)	56.0	13.5	52.9	15.1
Educational level of household heada	2.5	0.8	2.5	0.1
Sex of household head (1=male)	0.1	0.3	0.9	0.3
Hired labour (1=yes 0=no)	0.3	0.5	0.6	0.5
Household's HIV/AIDS status (1=affected, 0=non affected)	0.3	0.5	0.3	0.5
Extension contact (1=yes)	0.7	0.5	0.6	0.5
Morbidity (No. of ill household members)	1.5	0.8	1.5	0.6
Mortality (No. of adult deaths)	0.1	0.3	1.5	0.7
Natural capital				
Farm size (ha)	1.3	2.1	0.8	0.8
Land tenure (Ownership of title deed 1= yes)	0.8	0.4	0.5	0.5
Financial capital				
Household saving (1=yes)	0.6	0.5	0.4	0.5
Access to credit ((1=yes)	0.2	0.4	0.2	0.4
Remittances (1=yes)	0.3	0.5	0.2	0.4
Physical capital				
Total livestock value (KSh '000)	81.8	57.4	50.2	45.3
Total farm equipment (KSh '000)	79.0	14.4	28.8	56.1
Personal household items value (KSh '000)	27.4	87.7	20.2	95.3
Social capital				
Membership to group (1=yes)	0.9	0.3	0.8	0.4
Off-farm livelihood activities				
Formal employment (yes=1)	0.3	0.5	0.2	0.4
Non-formal employment (yes=1)	0.3	0.5	0.3	0.4
Trading (yes=1)	0.3	0.5	0.3	0.5
Service (yes=1)	0.1	0.3	0.3	0.5
Land renting out (yes=1)	1.9	0.3	2	0.3

Source: Household survey, 2004.

Dependent variables: adopters 1 = growing tissue-cultured banana, 0 = not growing tissue-cultured banana.

Table A1.2. Description and summary statistics of the variables used in the logistic regression analysis of farming households' continued use of tissue-cultured banana in Maragua (N=67).

	Continued use (N=48)		Dis-continue use (N=19)	
	Mean	SD	Mean	SD
Human capital				
Household size (persons)	4.8	2.7	4.8	2.7
Age of household head (years)	52.5	16.3	52.3	16.3
Educational level of household head	2.4	0.6	2.4	0.6
Sex of household head (1=male)	0.8	0.4	0.8	0.4
Hired labour (1=yes 0=no)	0.7	0.5	0.7	0.5
Household's HIV/AIDS status (1=affected, 0=non affected)	0.4	0.5	0.2	0.4
Extension contact (1=yes)	0.8	0.4	0.4	0.5
Morbidity (No. of ill household members)	0.1	0.3	0.2	0.4
Mortality (No. of adult deaths)	1.5	0.8	1.6	0.7
Natural capital				
Farm size (ha)	1.5	2.4	0.9	0.6
Land tenure (Ownership of title deed 1= yes)	0.8	0.4	0.6	0.5
Financial capital				
Household saving (1=yes)	0.6	0.5	0.5	0.5
Access to credit ((1=yes)	0.3	0.4	0.2	0.4
Remittances (1=yes)	0.3	0.5	0.3	0.5
Physical capital				
Total farm equipment (Ksh '000)	92.4	60.2	29.0	37.9
Social capital				
Membership to group (1=yes)	0.9	0.4	0.9	0.3
Off-farm livelihood activities				
Formal employment (yes=1)	0.3	0.5	0.3	0.5
Non-formal employment (yes=1)	0.4	0.4	0.3	0.5
Trading (yes=1)	0.3	0.5	0.3	0.5
Service (yes=1)	0.1	0.3	0.2	0.4
Renting (yes=1)	1.9	0.4	1.9	0.3

Source: Household survey, 2004.
Dependent variables: continued use: 1 = continued use of tissue-cultured plantlets, 2 = discontinued use of tissue-cultured plantlets.

Summary

Agricultural technology is known to be a catalyst for agricultural development and rural poverty reduction through increases in food production and/or reduction in production costs. Various government policy and strategic documents emphasize the important role of the agricultural sector as the leading driver for development in Kenya. The recently launched Economic Recovery Strategy for Wealth and Employment Creation (ERS Republic of Kenya 2003) identifies the adoption of appropriate agricultural technologies and practices as one of the imperative themes that will generate the surplus needed to feed the increasing population and to propel economic growth. However, the role of agricultural technology in poverty reduction is currently being played out in an increasingly multifaceted environment, featuring the growing complexity of farming rural households' livelihood strategies and the effects of HIV/AIDS on household assets and livelihood strategies. The current challenge facing agricultural research in countries threatened by HIV/AIDS is to develop technologies that meet the evolving challenges of HIV/AIDS-affected farming households without compromising productivity and sustainability of their livelihoods.

This study investigated the suitability of the tissue-cultured banana technology in rural farming households in Central Kenya in the context of HIV/AIDS. The study adopts a livelihood approach and provides detailed information on farming household assets and livelihood strategies. It examines the effects of HIV/AIDS on household assets and compares the consequent livelihood strategies undertaken in HIV/AIDS-affected and non-affected farming households. The study evaluates in detail banana farming as one of the livelihood strategies and assesses the significance of the tissue-cultured banana technology for the livelihood of the farming households. The effects of HIV/AIDS on the local extension services and how this influences the adoption of tissue cultured-banana technology adoption is investigated. A gender perspective of access to assets, HIV/AIDS impacts, livelihood activities and outcomes is integrated. The sample population was selected on the basis of use or non-use of the tissue-cultured banana among banana farming households. Within each of these samples, both HIV/AIDS-affected and non-affected households were selected. The data used in the analysis were collected through a mixed-method approach incorporating quantitative and qualitative data. The quantitative data was collected through a formal survey whereas qualitative data was collected through in-depth interviews, focus group discussions and key informant interviews. The study was conducted in Maragua district, which is a banana-growing region in Central Kenya where the tissue-cultured banana has been introduced. The study elaborates the following research questions:

1. How do assets available to farming households influence their livelihood options, activities and outcomes?

2. What are the impacts of HIV/AIDS on farming households' assets and consequently livelihood options, activities and outcomes?
3. What is the role of the tissue-cultured banana in the livelihood activities and outcomes of farming households?

Because farming households differ in their asset endowment, they undertake differing livelihood strategies and respond differently to both shocks and development interventions. To capture this heterogeneity, by means of factor and cluster analysis, a typology according to resource endowment was constructed yielding three types of farming households: low-, medium- and high-resource endowment farming households using factor and cluster analysis. The classification was based on households' asset endowment in terms of human capital (age, sex, education level of the household head and household size); natural capital (size of land); financial capital (savings and access to credit); physical capital (livestock, farm equipment and personal household item value); and social capital (membership in community organization).

The low resource endowment category is composed of a high proportion of elderly (65+) female-headed households with a low level of education of the household head and a high demographic dependency ratio. Households in this category have small pieces of land inherited from their parents, and half of them have no title deeds for the land. The households have no access to formal credit and no savings. They have low physical capital in terms of livestock, farm equipment and household items. They are producing mainly for own consumption and earn extra income by selling their own labour in agricultural activities.

The medium resource endowment group is composed almost entirely of male-headed households who have relative high education level. They have large pieces of land (on average >0.8 ha) with title deeds, have some savings and access to both formal and informal credit. They have more livestock, farm equipment and personal household items than the type one households. They mostly produce for own consumption and also for the market. They also earn extra income through formal and non-formal employment and are engaged in trading, service provision. They receive remittances from migrant household members.

The high resource endowment farming households have of educated male household heads with large land size which is registered with title deeds. They have access to informal credit but do not have savings. They have the highest livestock and farm equipment value. The majority of households in this group do not take on other income diversification activities and are mostly engaged in production of banana for the market and maize and beans for home consumption. They have the largest proportion of migrants and receive more remittances than the other two categories of households.

Generally, there is a high level of social capital in the study area as indicated by the large number of households involved in community organizations. The community groups existing in the area are: farming groups, rotating savings-and-credit associations (ROSCAs), burial societies, village committees, clan-based groups, religious groups and HIV/AIDS-associated groups. In general, households in medium and high resource endowment categories have higher number of household members engaged in groups than in the low resource endowment category. However, these groups may not necessarily lead to economic prosperity as most of them are homogenous and hardly connected to outside resources.

Despite the differences in assets endowment among the farming households, no significant differences were observed in their incidence of HIV/AIDS. However, differences were observed in asset endowment and household characteristics between HIV/AIDS-affected and non-affected farming households. Comparing affected and non-affected farming households showed that HIV/AIDS-affected households are mostly female-headed, have a significantly higher dependency ratio, and experience a greater shortage of labour despite their larger household size. Land sale, a commonly quoted strategy for household labour loss related to HIV/AIDS is rare in the study area. This is possibly due to cultural beliefs associated with land which deter the sale of ancestral land. Lack of transferability rights (as more than half of the households in sample population have no title deeds to their land) could also be a deterrence factor of land sale. Affected households have adopted various labour coping strategies such as labour re-allocation, hiring labour and bringing in relatives. Some affected households have stopped growing labour-intensive vegetable crops while some have altogether abandoned their land. To cater for the high direct and indirect costs, HIV/AIDS-affected households have various sources of income. These include borrowing from informal sources (ROSCAS, relatives and extended family members) and they also receive remittances from migrant household members. Membership in ROSCAS is crucial as it enables households to access informal credit. In addition to farming, HIV/AIDS-affected households have diversified income sources in the form of rent from land and they engage in labour migration.

The adoption of tissue-cultured banana is positively related to financial and physical capitals. The high and medium resource endowment households that have financial capital in form of savings and access to credit are more likely to grow tissue-cultured banana. This is primarily because of the high cost of the tissue-cultured plantlets. Furthermore, the results suggest that affected households are most likely not to adopt the tissue-cultured banana technology. Thus, for the technology to benefit resource-poor farming households who constitute the majority of the sample population, efforts must be made to increase their access to financial capital or lower the cost of tissue-cultured plantlets.

Increased farm output resulting from technology adoption is expected to result in various outcomes such as increased household food security, increased income through the sale of extra produce, and reinvestment into household activities. Both HIV/AIDS-affected and non-affected households growing tissue-cultured banana reported an increase in banana production, income and food, which they attributed to the adoption of the technology. However, HIV/AIDS-affected reported higher increases in production, income and food supply as a result of growing tissue-cultured banana.

Although the area extension services have not directly suffered the attrition caused by HIV/AIDS, they are lacking skills, finances and human capacity in providing appropriate services to farming households and in particular the HIV/AIDS-affected households. Not only will more extension workers be needed, but contents of extension services also need to change in order to be responsive to the AIDS epidemic. Extension services need to cater to the knowledge needs of women, the elderly and the young.

Female-headed households were found to be disadvantaged in several ways. The majority of them are in the low resource endowment farming household category. Firstly, female household heads have significantly lower education level than male-headed households. Secondly, their financial capital base is also low because they have no savings and their access to informal credit provided by community organisation is limited, having few household members engaged in these groups. In addition, female-headed households have significantly lower physical capital in terms of household assets which they could sell and get cash in times of crisis. Slightly over half of the female-headed households are HIV/AIDS-affected. These households have a significantly higher dependency ratio and a higher incidence of labour shortage than non-affected households despite their larger size. Renting out of land a strategy undertaken by HIV/AIDS-affected households is not common in female-headed households due to lack of transferability rights. Some of the household members have therefore opted to migrate to urban areas in search of employment, a strategy that further puts the households at risk of HIV/AIDS.

In conclusion, policy-makers and development agents targeting agricultural technology development for food security and poverty reduction should take into account the diversity of farming households. Farming household capabilities, assets and activities should play a major role in shaping policy on agricultural technology development. Labour-saving technologies may indeed be appropriate for many households, especially female-headed HIV/AID-affected households that lack cash for hiring labour. However, agricultural research should focus on developing low risk technologies in terms of financial requirement that can assure farming household food security, as well as cash income to pay for school fees and basic necessities.

Curriculum vitae

Faith Njeri Nguthi (1961, Kenya) graduated from the University of Nairobi with a BSc degree in Agriculture in 1985. She worked for the Ministry of Agriculture from 1985-1987 as an extension officer in Machakos district. From 1987-1989, she worked with the Kenya Agricultural Researcher Institute (KARI) based at the National Horticultural Research Centre (NHRC), Thika. In 1989, she was awarded a CIDA scholarship and studied for an MSc in Horticulture at the University of Nairobi. From 1994 to April 2003 she worked at NHRC, Thika, as the head of fruit section at the Centre from 1996-2003, the national team leader of the project 'Biotechnology to Benefit Small-holder Banana Producers in Kenya' from 1998-2002 and as the Deputy Centre Director from 1997-2003.

In 2003, Faith was awarded a PhD scholarship by the collaborative program between Africa Women Leaders in Agriculture and Environment (AWLAE), Winrock International (WI), and Wageningen University and Research Centre (WUR). The project was funded by Netherlands Directorate-General for International Cooperation (DGIS). Her research focused on the adoption of agricultural technologies in the context of HIV/AIDS.

She is currently working with the Kenya Agricultural Research Institute as a Senior Agricultural Research Officer.

Curriculum vitae

www.ingramcontent.com/pod-product-compliance
Lightning Source LLC
Chambersburg PA
CBHW061209240926
51846CB00006B/1167
* 9 7 8 9 0 8 6 8 6 0 6 9 2 *